Climate Change and Environment

Climate Change and Environment

DR. R. KAILASH

AMIGA PRESS. INC
Delhi -110051

AMIGA PRESS. INC
H.O :- 28-A, Indra Park, Chander Nagar
Delhi :- 110051 (India)
Tel :- 011-22504028, +91-98736-57000
Email :- amigapress@gmail.com

This edition is published by Amiga Press.Inc

ISBN : 978-81-929542-3-3

First Published 2018

PRINTED : AMIGA PRESS INDIA

Preface

Climate changes in response to changes in the global energy balance. On the broadest scale, the rate at which energy is received from the sun and the rate at which it is lost to space determine the equilibrium temperature and climate of Earth. This energy is then distributed around the globe by winds, ocean currents, and other mechanisms to affect the climates of different regions.

Factors that can shape climate are called climate forcings or "forcing mechanisms". These include such processes as variations in solar radiation, deviations in the Earth's orbit, mountain-building and continental drift, and changes in greenhouse gas concentrations. There are a variety of climate change feedbacks that can either amplify or diminish the initial forcing.

Some parts of the climate system, such as the oceans and ice caps, respond slowly in reaction to climate forcings, while others respond more quickly. Forcing mechanisms can be either "internal" or "external". Internal forcing mechanisms are natural processes within the climate system itself (e.g., the meridional overturning circulation). External forcing mechanisms can be either natural (e.g., changes in solar output) or anthropogenic (e.g., increased emissions of greenhouse gases).

The environment and the economy are obviously very closely related. This link is more than a mere principle; it is a necessity for sustainable development. Various economic tools and policies may promote sustainable development, or at least lead to a more environmentally conscious use of resources.

These tools or policies, such as the polluter-payer or consumer-payer approach, may be applied equally to producers, consumers and taxpayers and to enable the market to determine the correct overall cost of using resources. In many instances, however, for the actual value of natural resources to be taken into account, producers and economic agents need to change their attitudes.

Environmental science is an interdisciplinary academic field that integrates physical and biological sciences, (including physics, chemistry, biology, soil science, geology, and geography) to the study of the environment, and the solution of environmental problems. Environmental science provides an integrated, quantitative, and interdisciplinary approach to the study of environmental systems. Related areas of study include environmental studies and environmental engineering. Environmental studies incor- porate more of the social sciences for understanding human relationships, perceptions and policies towards the environment. Environmental engineering focuses on design and technology for improving environmental quality.

This book provides deep insight to various dimensions of issues relating to the subject.

—*Editor*

Contents

1

Climate Change

Climate change is a long-term change in the statistical distribution of weather patterns over periods of time that range from decades to millions of years. It may be a change in the average weather conditions or a change in the distribution of weather events with respect to an average, for example, greater or fewer extreme weather events. Climate change may be limited to a specific region, or may occur across the whole Earth. In recent usage, especially in the context of environmental policy, climate change usually refers to changes in modern climate. It may be qualified as anthropogenic climate change, more generally known as global warming or anthropogenic global warming (AGW).

ISSUE OF CLIMATE CHANGE

The issue of climate change detection introduces the wider concern of future climate change. Although predictions of climate change thousand, millions and even hundreds of millions of years from now could be made based on our knowledge of longer term climate changes, I will be concerned only with the current decadal to century time scale. Clearly, prediction of climate change over the next 100 to 150 years is based solely on model simulations.

Understandably, the vast majority of modelling has concentrated on the effects of continued anthropogenic pollution of the atmosphere by greenhouse gases, and to a lesser extent, atmospheric aerosols. The main concern, at present, is to determine how much the Earth will warm in the near future.

GCM Climate Simulations

During the last decade or so, a number of complex GCMs have attempted to simulate future anthropogenic climate change. Earlier models studied equilibrium climate change associated between two, presumably stable, states of climate. The results of these are discussed in the first IPCC report. Associated with a doubling of pre-industrial atmospheric CO_2, the following conclusions have been made:

- A global average warming at or near the Earth's surface of between 1.5 and 4.5°C, with a "best guess" of 2.5°C, will occur;
- The stratosphere will experience a significant cooling;
- Surface warming will be greater at high latitudes in winter, but less during the summer;
- Global precipitation will increase by 3 to 15%;
- Year-round increases in precipitation in high-latitude regions are expected, whilst some tropical areas may experience small decreases.

More recent time-dependent GCMs which couple the atmospheric and oceanic components of the climate system together, provide more reliable estimates of greenhouse-gas-induced climate change. For steadily increasing greenhouse forcing, the global rise in temperature is typically less than the equilibrium rise corresponding to an instantaneous forcing. Significant results indicate:

- A global average warming of 0.3°C per decade, assuming non-interventionist greenhouse gas emission scenarios;
- A natural variability of about 0.3°C in global surface air temperature on decadal time scales;
- Regional patterns of temperature and precipitation change similar to equilibrium experiments, although warming is reduced in high latitude oceans where deep water is formed.

The ability to model the time-dependent nature of the climate system more adequately has allowed scientists to investigate the damping effects of the oceans on climate change. Because the response time of the oceans, in particular the deep ocean, is much longer than for the free atmosphere, they have a regulating or delaying effect on the warming associated with enhanced greenhouse forcing. In addition, the transient GCMs have allowed increased attention to focus on the critical role of feedback processes in determining the climate's response to forcing perturbations.

Greenhouse Feedbacks

GCMs have estimated the CO_2 doubling temperature change in the absence of feedback processes to be approximately 1.2°C. The existence of feedback loops within the climate system results in a climate sensitivity of 1.5°C to 4.5°C, and makes an otherwise linear association between radiative forcing and global temperature distinctly non-linear. Three of the most important direct climatic feedbacks to greenhouse forcing include the water vapour feedback, the cloud feedback and the ice-albedo feedback.

Water Vapour Feedback

The importance of water vapour feedback in climate change has long been recognised. The concentration of water vapour in the atmosphere increases rapidly with rising temperature (about 6%/°C); this is the basis for the strong positive water vapour feedback seen in current climate models (whereby increases in temperature produce increases in atmospheric water vapour which in turn enhance the greenhouse forcing leading to further warming). All current GCMs simulate a strong positive water vapour feedback and are in agreement with observational data.

Cloud Feedback

Cloud feedback is the term used to encompass effects of changes in cloud and their associated radiative properties on a change of climate, and has been identified as a major source of uncertainty in climate models. This feedback mechanism incorporates both changes in cloud distribution (both horizontal and vertical) and changes in cloud radiative properties (cloud optical depth and cloud droplet distribution); these are not mutually independent. Although clouds contribute to the greenhouse warming of the climate system by absorbing more outgoing infrared radiation (positive feedback), they also produce a cooling through the reflection and reduction in absorption of solar radiation (negative feedback). It is generally assumed that low clouds become more reflective as temperatures increase, thereby introducing a negative feedback, whilst the feedback from high clouds depends upon their height and coverage and could be of either sign.

Ice-Albedo Feedback

The conventional explanation of the amplification of global warming by snow and ice feedback is that a warmer Earth will have less snow cover, resulting in a lower global albedo and consequent absorption of more radiation, which in turn causes a further warming of the climate. Most GCMs 1990, 1991) have simulated this positive surface albedo feedback, but significant uncertainties exist over the size of the effect, particularly with sea-ice.

Greenhouse Gas Feedbacks

In addition to these climatic feedbacks, there are other non-climatic feedbacks which may enhance or diminish the increase in atmospheric greenhouse gas accumulation. Potentially, there are numerous feedback processes which may act on the carbon cycle, thus affecting the transfer of carbon dioxide between the various components. Carbon storage in the oceans will be affected by both changes in water temperature and ocean circulation resulting from global warming. Warmer water stores less CO_2, and future atmospheric CO_2 increase may be amplified by something like 5 to 10% by this effect. A slow down in vertical overturning within the oceans will reduce the uptake of atmospheric CO_2. In addition, reduced nutrient recycling would limit primary biological productivity in the surface oceans, thereby increasing further the surface water partial pressure of CO_2, and consequently atmospheric CO_2.

Carbon cycle feedbacks involving terrestrial carbon storage may include carbon dioxide fertilisation of plant photosynthesis (a negative feedback); eutrophication involving nitrogen and phosphate nutrient fertilisation (a negative feedback); temperature feedbacks on the length of growing season and photosynthesis/respiration rates (both positive and negative feedbacks) and changes in the geographical distribution of vegetation. The methane gas cycle may also experience feedback effects. Recent attention has been focused on northern wetlands and permafrost, where increases in temperature and soil moisture would result in significant increases in methane release into the atmosphere.

Despite the increased ability of transient GCMs to model climate feedback processes, there is no compelling evidence to warrant changing the equilibrium sensitivity to doubled CO_2 from the range of 1.5 to 4.5°C as estimated by IPCC, 1990a.

Impacts of Future Climate Change

If the climate system responds to enhanced greenhouse forcing in a manner that has been projected by climate models, the Earth is likely to experience a number of detrimental impacts.

Obviously, much depends upon the magnitude of the climate sensitivity, a variable as yet poorly constrained by the present understanding of the climate system.

In addition, climate change, and therefore climate impacts, are poorly defined at the regional level. Nevertheless, it is possible to make generalisations of the likely consequences of future anthropogenic climate change, and to attach levels of certainty to these. The IPCC Climate Change Impacts Assessment provides an excellent review in this context.

As the climate system shifts to a new equilibrium, each of its components and sub-components will respond to the complex interaction of feedback loops.

Climate impacts can therefore be expected to occur throughout the system. Some of the major impact areas identified include agriculture, forestry, natural ecosystems, hydrology and water resources, human settlements and health, and oceans and coastal zones. Global-mean surface temperature increase, regional temperature increases, precipitation increases and decreases, soil moisture availability, climatic variability and the occurrence of extreme events such as hurricanes are all likely to influence the nature of these impacts. Climatic impacts, whether detrimental or benign, will also have to be viewed in the context of a rapidly expanding global population.

Agriculture

There is no overall consensus as to whether anthropogenic climate change will increase or decrease global agricultural potential. Negative impacts could be experienced at the regional level as a result of changes in weather, moisture availability, pests associated with climate change and changes in ground level ozone associated with pollutants.

Areas of existing vulnerability, including SE Asia, the Sahel in N. Africa and parts of South America, may experience the most severe effects. Parts of the high and mid latitudes may see increases in potential productivity as the length of the growing season expands, but this is likely to be confined largely to the Northern Hemisphere.

On balance, global food production may be maintained at or near present day levels; however, the cost of achieving this is unclear, especially in view of a rapidly growing world population.

Forestry

The rotation period of forests is long and trees will find it increasingly difficult to adjust to the swifter changes of climatic regimes. Actual impacts will depend upon the physiological adaptability of trees and their host-parasite relationships. Tree species close to their biological limits (in terms of temperature and moisture) will be most sensitive to climate change.

Areas experiencing warmer, drier climates may also see an increase in the incidence of forest-fires. As well as the impact to forest ecology, climate change will also augment social stresses of forest-dependent populations, and consequent anthropogenic damages to forests can be expected. Such increased and non-sustainable forest use will put further pressure on forest investments, conservation and management.

Natural Terrestrial Ecosystems

Most climate models agree that a reduction in the equator to pole temperature gradient, at least on land, will be experienced in a greenhouse world.

Consequently, climatic zones could shift several hundred kilometres towards the poles during the next century, although the rate of transition is uncertain. As flora and fauna would lag behind these climatic shifts, they would be subject to considerable stresses. Many species may be unable to adapt to different climatic and meteorological regimes, and will become extinct. The rate of projected climate changes is critical in determining the nature of impacts on terrestrial ecosystems and their ability to adapt. Even a projected temperature increase of 0.2°C per decade is at least twice the maximum rate of increase that most ecosystems could withstand. The most susceptible ecological systems will include montane, alpine, polar, island and coastal communities where climatic changes will add to existing stresses.

Hydrology and Water Resources

Relatively small climatic variations can create large problems for water resources, especially in arid and semi-arid regions. However, the climatic impacts on hydrology at the regional scale are uncertain, and will be influenced by a complex mix of temperature, precipitation, evaporation, soil moisture availability and run off changes.

For many marginal areas, such as the Sahelian region in Africa, the rates of present desertification are likely to increase, having significant implications for agriculture, water storage, distribution and power generation. Other areas, such as NW Europe, may experience increases in precipitation, with resultant increases in the risk of flooding, as water tables rise. Patterns of water demand will need to alter in response to changes in water supply. This will be true for areas at risk from both drought and flooding, and effective water management strategies will be required.

Oceans and Coastal Zones

Global warming will accelerate sea-level rises due to thermal expansion of seawater and the melting of land-based ice sheets. If current GCMs projections of a 4cm per decade rise are reasonable, sea level may stand at least half a metre higher by the end of the next century. This will seriously threaten many low-lying islands and coastal zones, rendering some countries uninhabitable. Other impacts include the increased risk of coastal flooding and salination of fresh groundwater supplies, exacerbated by an increasing occurrence of droughts or storms. Rapid sea-level rise would damage the coastal ecology, threatening important fisheries. Changes in ocean circulation are also projected to occur as a result of changing global climate. This will not only influence marine ecosystems (due to changes in upwelling zones), but will dramatically affect regional meteorological patterns (in response to the shifting heat balance). Palaeoclimatic studies have revealed that rapid circulation changes, within only a few decades, have occurred in the past.

Human Settlements and Health

The most vulnerable populations to the impacts of climate change will be those already under pressure from social, economic and existing climate stresses. These will include those in developing countries, in the lower income groups, residents of coastal lowlands and islands, populations in semi-arid grasslands, and the urban poor. Increased exposure to natural hazards, such as coastal or river flooding, drought, landslides, storms and hurricanes, will prove detrimental to those most at risk. In the industrialised world many of the largest cities are situated in low-lying regions, and rises in sea level will threaten large land areas of immense economic wealth.

Major health impacts can be expected as changes in precipitation and temperature will affect the patterns of vector-borne and viral diseases.

Incidences of malaria, for example, could spread to higher latitudes outside pre-existing tropical bounds. Such impacts could initiate large migrations of people, causing substantial disruptions of settlement patterns and social instability.

TIME SCALE OF CLIMATIC CHANGE

The importance of considering different time scales when investigating climate change has already been identified. Climate varies on all time scales, in response to random and periodic forcing factors. Across all time periods from a few years to hundreds of millions of years there is a white (background) noise of random variations of the climate, caused by internal processes and associated feedback mechanisms, often referred to as stochastic or random mechanisms. Such randomness accounts for much of the climate variation, and owes its existence to the complex and chaotic behaviour of the climate system in responding to forcing. An essential corollary of the existence of random processes is that a large proportion of climate variation cannot be predicted.

Of far more relevance are the periodic forcing factors, for by understanding their mechanisms and the impacts they have on the global climate, it is possible to predict future climate change. How the climate system responds to periodic forcing factors, however, is often not clear. If it is assumed that the climate system responds in a linear fashion to periodic forcing, variations in climate will exhibit similar periodicity. If, however, the response of the system to forcing is strongly non-linear, the periodicities in the response will not necessarily be identical to the periodicities in the forcing factor(s). Frequently, the climate responds in a fashion intermediate between the two.

There are many climate forcing factors spanning an enormous range of periodicities. The longest, 200 to 500 million years, involves the passage of our Solar System through the galaxy, and the variations in galactic dust. These may be considered to be external forcing mechanisms. Other long time scale variations (10^6 to 10^8 years) include the non-radiative forcing mechanisms, such as continental drift, orogeny (mountain building) and isostasy (vertical movements in the Earth's crust affecting sea level). The response of the climate system to this combination of forcing factors itself depends upon the different response times of the various components of the system. The overall climatic response will then be determined by the interactions between the components. The atmosphere, surface snow and ice, and surface vegetation typically respond to climatic forcing over a period of hours to days. The surface ocean has a response time measured in years, whilst the deep ocean and mountain glaciers vary only over a period spanning hundreds of years. Large ice sheets advance and withdraw over thousands of years whilst parts of the geosphere (e.g. continental weathering of rocks) respond only to forcing periods lasting hundreds of thousands to millions of years.

The response of the climate system to episodes of forcing can be viewed as a form of resonance. When the time period of forcing matches most closely the response time of a particular system component, the climatic response will be greatest within that component.

Milankovitch forcing, for example, with periods of tens of thousands of years will be manifest in the response of the ice sheets, and the overall response of the climate system will be dominated by changes within the cryosphere. In addition, longer response times of certain components of the climate system modulate, through feedback processes, the short term responses. The response of the deep ocean to short term forcing (e.g. enhanced greenhouse effect, solar variations, for example, will tend to attenuate or smooth the response of the atmosphere.

Throughout the remainder, it should be recognised that a range of time scales applies to climate forcing mechanisms, radiative and non-radiative, external and internal, and to the response of the different components of the climate system.

Climate Sensitivity

The concept of feedback is related to the climate sensitivity or climate stability. It is useful to have a measure of the strength of various feedback processes which determine the ultimate response of the climate system to any change in radiative forcing. In general terms, an initial change in temperature due to a change in radiative forcing, $\Delta T_{forcing}$, is modified by the complex combination of feedback processes such that:

$$\Delta T_{final} = \Delta T_{forcing} + \Delta T_{feedback}$$

where $\Delta T_{feedback}$ is the temperature change resulting from feedback and ΔT_{final} is the overall change in temperature between the initial and final equilibrium states. The degree to which feedback processes influence the final climatic response is a measure of the sensitivity of the climate system.

Equation can be rewritten as follows:

$$\Delta T_{final} = f\Delta T_{forcing}$$

where f is called the feedback factor. When only one feedback mechanism is operative the solution to equation 6 is simple, assuming both f and $\Delta T_{forcing}$ are known. When more than one feedback is operative, matters become more complicated. For two feedbacks, the net effect is given by:

$$f = f_1 f2/ (f_1 + f2 - f_1 f2)$$

where f_1 and f2 are the feedback factors of the two feedback processes. Clearly, the feedback factors are neither additive nor multiplicative. A feedback operating alone with a factor of 2 would double the initial climatic response to forcing. If a second feedback with factor 1.5 acted with it, the overall feedback would be enhanced by a factor of 6. It can be seen, then, that a combination of feedback processes could dramatically affect the climate as it responds to only a small change in radiative forcing.

The climate's sensitivity can be mathematically determined in another way. From satellite observations, it has been shown that changes in global temperature are approximately proportional to changes in radiative forcing. If we assume an instantaneous + change in climate, from one equilibrium state to another, then:

$$\Delta Q = \lambda \Delta T$$

where ΔQ is change in radiative forcing (expressed in terms of the net downward radiative flux at the top of the troposphere), ΔT is the change in global temperature, and λ is a measure of the climate sensitivity.

Based on equation, the climate sensitivity is usually expressed in terms of the temperature change associated with a specified change in radiative forcing, usually a doubling of the atmospheric carbon dioxide content. Thus, the carbon dioxide doubling temperature, ΔT_{2x}, is given by,

$$\Delta T_{2x} = \Delta Q_{2x}/\lambda$$

where ΔQ_{2x} is 4.2 Wm^{-2}. The magnitude (and sign) of ΔT_{2x} will depend on λ, the climate sensitivity, which is determined by the net affect of the climate feedback processes. Despite extensive climate modelling over the last 2 decades to understand the problem of contemporary global, it is this parametre that is proving hard to define numerically.

As explained earlier, the idea of static equilibrium and instantaneous climatic response represents an unrealistic situation. To take into account the dynamic and transient nature of the climate's response to forcing, a more complicated equation linking ΔT and ΔQ is required if the evolution of the response with time is to be determined.

In the case:

$$\Delta Q = \lambda \Delta T + C d\Delta T/\delta t$$

Here, the change in radiative forcing, DQ, is balanced by:

- The change in the outgoing radiative flux at the tropopause caused by the response of the climate system including feedback; and
- The energy stored in the system, $C\delta\Delta T/\delta t$, where C is the system's heat capacity and t is time.

The latter term in equation simulates the time-dependent nature of the climate system's response. The main contributor to the system's heat capacity is the world's oceans. The heat capacity of water is large compared to that of air and is therefore able to store much more energy. Additionally, the high heat capacity means that the oceans take time to heat up (or cool down) and hence slow the surface temperature response to any change in radiative forcing: the transient response will always be less than the equilibrium response.

Solution of Equation leads to the definition of the response time of the climate system, τ, such that:

$$\tau = C/\lambda$$

If the heat capacity of the climate system is large, the response time is large. Equally, if the climate sensitivity is small the response time is large. λ is also known as the radiative damping coefficient. Here, analogy is made between the response of the climate system and an oscillating spring. If a spring has a high damping coefficient, it will stop oscillating soon after it has been set in motion. Similarly, if the radiative damping coefficient is large, the climate will respond quickly and τ will be small.

CLIMATE SYSTEM

The key to understanding global climate change is to first understand what global climate is, and how it operates. At the planetary scale, the global climate is regulated by how much energy the Earth receives from the Sun. However, the global climate is also affected by other flows of energy which take place within the climate system itself. This global climate system is made up of the atmosphere, the oceans, the ice sheets (cryosphere), living organisms (biosphere) and the soils, sediments and rocks (geosphere), which all affect, to a greater or less extent, the movement of heat around the Earth's surface.

The atmosphere plays a crucial role in the regulation of Earth's climate. The atmosphere is a mixture of different gases and aerosols (suspended liquid and solid particles) collectively known as air. Air consists mostly of nitrogen (78%) and oxygen (21%). However, despite their relative scarcity, the so-called greenhouse gases, including carbon dioxide and methane, have a dramatic effect on the amount of energy that is stored within the atmosphere, and consequently the Earth's climate. These greenhouse gases trap heat within the lower atmosphere that is trying to escape to space, and in doing so, make the surface of the Earth hotter. This heat trapping is called the natural greenhouse effect, and keeps the Earth 33°C warmer than it would otherwise be. In the last 200 years, man-made emissions of greenhouse gases have enhanced the natural greenhouse effect, which may be causing global warming.

The atmosphere however, does not operate as an isolated system. Flows of energy take place between the atmosphere and the other parts of the climate system, most significantly the world's oceans.

For example, ocean currents move heat from warm equatorial latitudes to colder polar latitudes. Heat is also transferred via moisture. Water evaporating from the surface of the oceans stores heat which is subsequently released when the vapour condenses to form clouds and rain. The significance of the oceans is that they store a much greater quantity of heat than the atmosphere. The top 200 metres of the world's oceans store 30 times as much heat as the atmosphere. Therefore, flows of energy between the oceans and the atmosphere can have dramatic effects on the global climate.

The world's ice sheets, glaciers and sea ice, collectively known as the cryosphere, have a significant impact on the Earth's climate. The cryosphere includes Antarctica, the Arctic Ocean, Greenland, Northern Canada, Northern Siberia and most of the high mountain ranges throughout the world, where sub-zero temperatures persist throughout the year. Snow and ice, being white, reflect a lot of sunlight, instead of absorbing it. Without the cryosphere, more energy would be absorbed at the Earth's surface rather than reflected, and consequently the temperature of the atmosphere would be much higher.

All land plants make food from the photosynthesis of carbon dioxide and water in the presence of sunlight. Through this utilisation of carbon dioxide in the atmosphere, plants have the ability to regulate the global climate. In the oceans, microscopic plankton utilise carbon dioxide dissolved in seawater for photosynthesis and the manufacture of their tiny carbonate shells. The oceans replace the utilised carbon dioxide by "sucking" down the gas from the atmosphere. When the plankton die, their carbonate shells sink to the seafloor, effectively locking away the carbon dioxide from the atmosphere. Such a "biological pump" reduces by at least four-fold the atmospheric concentration of carbon dioxide, significantly weakening the Earth's natural greenhouse effect, and reducing the Earth's surface temperature.

CLIMATE MODELLING

Climate models attempt to simulate the behaviour of the climate system. The ultimate objective is to understand the key physical, chemical and biological processes which govern climate. Through understanding the climate system, it is possible to: obtain a clearer picture of past climates by comparison with empirical observation, and; predict future climate change. Models can be used to simulate climate on a variety of spatial and temporal scales.

Sometimes one may wish to study regional climates; at other times global-scale climate models, which simulate the climate of the entire planet, will be desired.

Henderson-Sellers and Robinson, Henderson-Sellers and McGuffie and Schneider provide detailed introductory discussions of the methods and techniques involved in climate modelling.

There are three major sets of processes which must be considered when constructing a climate model:

- *Radiative*: The transfer of radiation through the climate system (e.g. absorption, reflection);
- *Dynamic*: The horizontal and vertical transfer of energy (e.g. advection, convection, diffusion);
- *Surface process*: Inclusion of processes involving land/ocean/ice, and the effects of albedo, emissivity and surface-atmosphere energy exchanges.

These are the processes fundamental to the behaviour of the climate system that were discussed.

The basic laws and other relationships necessary to model the climate system are expressed as a series of equations. These equations may be empirical in derivation based on relationships observed in the real world, they may be primitive equations which represent theoretical relationships between variables, or they may be a combination of the two. Solving the equations is usually achieved by finite difference methods; it is therefore important to consider the model resolution, in both time and space i.e. the time step of the model and the horizontal/vertical scales.

Climate is the long-term statistical expression of short-term weather. Climate can be defined qualitatively as "expected weather" or quantitatively by statistical expressions such as central tendencies and variances. The overall distribution of climatological parametres, bounded by weather extremes, defines the climatic variability.

Changes in climate can be defined by the differences between average conditions at two separate times. Climate may vary in different ways and over different time scales. Variations may be periodic quasi-periodic or non-periodic.

This guide represents an up-to-date review of climate change. Throughout, the focus has essentially been on global climate change, although reference to regional scale climatic change has been made if and when necessary. On their own each is a broadly self-contained discussion of a specific sub-issue of importance.

The purpose of the guide is to serve as a first resort for undergraduates, applicable to first, second and third years in environmental sciences, geography and related disciplines. It is sufficient in depth and rigour for final year students but equally valuable as an information resource for those in their "pre-finals" years. It is not a complete guide but serves to review and illustrate the key factors of climate change over time and space. It contains an extensive list of references which the student is urged to consult.

The authors recognise that the study of global climate change is a rapidly developing discipline, and recent empirical evidence of climate change, and modelling evidence of the causes of climate change may have been omitted. In this sense, the guide represents a snapshot in time of the current understanding of climate change, which is open to future development. The opportunity therefore exists for readers of this guide to comment upon its content, both regarding improvements to the existing material and new ideas for additional entries. Comments and responses regarding the Global Climate Change Student Guide should be sent to The Global Climate Change Information Programme at the address on the front cover.

Simplifying the Climate System

All models must simplify what is a very complex climate system. This is in part due to the limited understanding that exists of the climate system, and partly the result of computational restraints. Simplification may be achieved in terms of spatial dimensionality, space and time resolution, or through parametreisation of the processes that are simulated. The simplest models are zero order in spatial dimension. The state of the climate system is defined by a single global average. Other models include an ever increasing dimensional complexity, from 1-D, 2-D and finally to 3-D models. Whatever the spatial dimension of a model, further simplification takes place in terms of spatial resolution. There will be a limited number of, for example, latitude bands in a 1-D model, and a limited number of gridpoints in a 2-D model. The time resolution of climate models varies substantially, from minutes to years depending on the nature of the models and the problem under investigation.

In order to preserve computational stability, spatial and temporal resolution have to be linked. This can pose serious problems when systems with different equilibrium time scales have to interact as a very different resolution in space and time may be needed.

Parametreisation involves the inclusion of a process as a simplified (sometimes semi-empirical) function rather than an explicit calculation from first principles. Subgridscale phenomena such as thunderstorms, for example, have to be parametreised as it is not possible to deal with these explicitly. Other processes may be parametreised to reduce the amount of computation required.

Certain processes may be omitted from the model if their contribution is negligible on the time scale of interest. For example, there is no need to consider the role of deep ocean circulation whilst modelling changes over time scales of years to decades. Some models may handle radiative transfers in great detail but neglect or parametreise horizontal energy transport. Other models may provide a 3-D representation but contain much less detailed radiative transfer information.

Given their stage of development, and the limitations imposed by incomplete understanding of the climate system and computational constraints, climate models cannot yet be considered as predictive tools of future climate change. They can, however, offer a valuable window on the workings of the climate system, and of the processes that have influenced both past and present climate.

Modelling the Climatic Response

The ultimate purpose of a model is to identify the likely response of the climate system to a change in any of the parametres and processes which control the state of the system. The climate response occurs in order to restore equilibrium within the climate system. For example, the climate system may be perturbed by the radiative forcing associated with an increase in carbon dioxide (a greenhouse gas) in the atmosphere. The aim of the model is then to assess how the climate system will respond to this perturbation, in an attempt to restore equilibrium.

The nature of the model can be one of two modes. In equilibrium mode, no account is taken of the energy storage processes which control the evolution of the climate response with time. It is assumed that the climate response occurs instantaneously following the system perturbation. The inclusion of energy storage processes allows the model to be run in transient mode, simulating the development of the climate response with time. Generally, both equilibrium and transient models will be run twice, once in a control run with no forcing, then over the test run including forcing and perturbation of the climate system. (For a further discussion of the concepts of equilibrium and transient climate responses, the reader may wish to refer back to.)

The climate sensitivity and the role of feedback are critical parametres whatever the model formulation. In the most complex models, the climate sensitivity will be calculated explicitly through simulations of the processes involved. In simpler models this factor is parametreised by reference to the range of values suggested by the more complex models. This approach, whereby more sophisticated models are nested in less complex models, is common in the field of climate modelling.

The Climate Models

It is often convenient to regard climate models as belonging to one of four main categories:

- Energy balance models (EBMs);
- One dimensional radiative-convective models (RCMs);
- Two-dimensional statistical-dynamical models (SDMs);
- Three-dimensional general circulation models (GCMs).

These models increase in complexity, from first to last, in the degree to which they simulate the particular processes and in their temporal and spatial resolution. The simplest models permit little interaction between the primary processes, radiation, dynamics and surface processes, whereas the most complex models are fully interactive.

It is not always necessary or beneficial to the analysis, however, to invariably choose the more sophisticated models. The choice of model depends upon the nature of the analysis For example, the simpler models, unlike the 3-D GCMs, may be run many times in sensitivity studies which test the influence of modelling assumptions.

For simulation experiments, which require complex modelling of the physical, chemical and biological processes inherent in the climate system, more sophisticated models may indeed be more appropriate. Computational cost is always an important factor to consider when choosing a climate model.

Energy Balance Models

Energy balance models (EBMs) simulate the two most fundamental processes governing the state of the climate:

- The global radiation balance (i.e. between incoming solar and outgoing terrestrial radiation)), and;
- The latitudinal (equator-to-pole) energy transfer.

EBMs are usually 0-D or 1-D in form. In 0-D EBMs, the Earth is considered as a single point in space, and in this case only the first process listed above is modelled. In 1-D models, the dimension included is latitude.

Temperature for each latitude band is calculated using the appropriate latitudinal value for the various climatic parametres (e.g. albedo, energy flux etc.).

Latitudinal energy transfer is usually estimated from a linear empirical relationship based on the difference between the latitudinal temperature and the global average temperature. Other factors that may be included in the model equation are the time-dependent energy storage and the energy flux into the deep ocean.

Radiative-Convective Models

Radiative-convective models (RCMs) are 1-D or 2-D, with height the dimension that is invariably present.

RCMs simulate in detail the transfer of energy through the depth of the atmosphere, including:

- The radiative transformations that occur as energy is absorbed, emitted and scattered, and;
- The role of convection, energy transfer via vertical atmospheric motion, in maintaining stability.

2-D RCMs also simulate horizontally-averaged energy transfers.

RCMs contain detailed information about the radiation streams or energy cascades - the fluxes of terrestrial and solar radiation - that occur throughout the depth of the atmosphere.

By considering parametres such as surface albedo, cloud amount and atmospheric turbidity, the heating rates of a number of atmospheric layers are calculated, based on the imbalance between the net radiation at the top and bottom of each layer.

If the calculated vertical temperature profile (lapse rate) exceeds some stability criterion (critical lapse rate), convection is assumed to take place (i.e. the vertical mixing of air) until the stability criterion is no longer breached.

This process is called the convective adjustment. As with EBMs, time-dependent calculations (energy storage and fluxes into the deep ocean) may be made, depending on the nature of the analysis. RCMs are most useful in studying forcing perturbations which have their origin within the atmosphere, such as the effects of volcanic pollution.

Statistical-Dynamical Models

Statistical-dynamical models (SDMs) are generally 2-D in form, with usually one horizontal and one vertical dimension, though variants with two horizontal dimensions have been developed. Standard SDMs combine the horizontal energy transfer modelled by EBMs with the radiative-convective approach of RCMs. However, the equator-to-pole energy transfer is simulated in a more sophisticated manner, based on theoretical and empirical relationships of the cellular flow between latitudes.

Parametres such as wind speed and wind direction are modelled by statistical relations whilst the laws of motion are used to obtain a measure of energy diffusion as in an EBM. Hence the description statistical-dynamical. They are particular useful in investigations of the role of horizontal energy transfer, and the processes which disturb that transfer directly.

General Circulation Models

General circulation models (GCMs) represent the most sophisticated attempt to simulate the climate system.

The 3-D model formulation is based on the fundamental laws of physics:

- Conservation of energy;
- Conservation of momentum;
- Conservation of mass, and;
- The Ideal Gas Law †

These are the same physical laws which formed the discussion of the climate system in theoretical terms. A series of primitive equations describing these laws are solved, resulting in an estimate of the wind field, which is expressed as a function of temperature. Processes such as cloud formation are also simulated.

To compute the basic atmospheric variables at each gridpoint requires the storage, retrieval, recalculation and re-storage of 10^5 figures at every time-step.

Since the models contain thousands of grid points †, GCMs are computationally expensive. However, being 3-D, they can provide a reasonably accurate representation of the planetary climate, and unlike simpler models, can simulate global and continental scale processes (e.g. the effects of mountain ranges on atmospheric circulation) in detail. Nevertheless, most GCMs are not able to simulate synoptic (regional) meteorological phenomena such as tropical storms, which play an important part in the latitudinal transfer of energy and momentum. The spatial resolution of GCMs is also limited in the vertical dimension. Consequently, many boundary layer processes must be parametreised.

To date, most GCMs only model the atmospheric component of the climate system, and have therefore been equilibrium in nature. Recently, a new breed of transient coupled atmosphere-ocean GCMs has evolved, in an attempt to simulate more accurately the climate system. Such coupled models create major computational difficulties because the atmosphere and ocean components respond over vastly different time scales.

In summary, GCMs can be considered to simulate reasonably accurately the global and continental-scale climate, but confidence is lacking in the regional detail.

Confidence and Validation

Although climate models should aid understanding in the processes which control and perturb the climate, the confidence placed in such models should always be questioned. Critically, it should be remembered that all climate models represent a simplification of the climate system, a system which indeed may ultimately prove to be too complex to model.

Given that many of the processes which are modelled occur over time scales so long that it is impossible to test model results against real-world observations, it is also arguable that climate modelling is, in some respects. philosophically suspect. Model performance can be tested through the simulations of shorter time scale processes but short-term performance may not necessarily reflect long-range accuracy.

Climate models must therefore be used with care and their results interpreted with due caution. Margins of uncertainty must be attached to any model projection.

Uncertainty margins can be derived by the comparison of the results of different model experiments or through sensitivity studies, in which key assumptions are altered to determine the role they play in influencing the final climatic response.

Validation of climate models (testing against real-world data) provides the only objective test of model performance. As far as GCMs are concerned, validation exercises have revealed a number of deficiencies in their simulations of present-day conditions:

- Modelled stratospheric temperatures tend to be lower than equivalent instrumental observations;
- Modelled mid-latitude westerlies tend to be too strong; easterlies are too weak;
- Modelled sub-polar low pressure systems in winter tend to be too deep and displaced east, and;
- Day-to-day variability is lower than in the real world.

Finally, it has been observed that some models suffer from climate drift. The background climate shifts as the simulation proceeds, despite the absence of any climate forcing.

Causes of Climate Change

The global climate must be viewed as operating within a complex atmosphere/earth/ocean/ice/land system. Any change to this system, resulting in climate change, is produced by forcing agents - the causes of climate change. Such forcing agents may be either internal or external.

External forcing mechanisms involve agents acting from outside the climate system. By contrast, internal mechanisms operate within the climate system itself. These are discussed separately. In addition, forcing mechanisms may be non-radiative or radiative.

The Atmosphere

The atmosphere is a mixture of different gases and aerosols (suspended liquid and solid particles) collectively known as air which envelopes the Earth, forming an integrated environmental (climate) system with all the Earth's components. The atmosphere provides various functions, not least the ability to sustain life. Of prime interest for a discussion on climate change, however, is its ability to control the Earth's energy budget. To understand this process, it will be necessary to study in more detail the composition of the atmosphere

The Composition of the Atmosphere

Let us consider first the atmospheric gases. illustrates the average gaseous composition of dry air below 25km. Although traces of atmospheric gases have been detected well out into space, 99% of the mass of the atmosphere lies below about 25 to 30km altitude, whilst 50% is concentrated in the lowest 5km (less than the height of Mount Everest).

Table. Average Composition of the Atmosphere below 25km

Component	*Chemical Abbreviation*	*Volume% (dry air)*
Nitrogen	N_2	78.08
Oxygen	O_2	20.98
Argon‡	Ar	0.93
Carbon dioxide	CO_2	0.035
Neon‡	Ne	0.0018
Helium‡	He	0.0005
Hydrogen	H	0.00006
Krypton‡	Kr	0.0011
Xenon‡	Xe	0.00009
Methane	CH_4	0.0017
Ozone†	O_3	0.00006

† Strictly speaking, the concentration of ozone in the atmosphere is variable.

‡ Inert gases.

This gaseous mixture remains remarkably uniform in composition, and is the result of efficient biogeochemical recycling processes and turbulent mixing in the atmosphere. The two most abundant gases are nitrogen (78% by volume) and oxygen (21% by volume), and together they make up over

99% of the lower atmosphere. There is no evidence that the relative levels of these two gases are changing significantly over time.

Despite their relative scarcity, the so-called greenhouse gases play an important role in the regulation of the atmosphere's energy budget.

Carbon dioxide

Carbon dioxide (CO_2), the most important of these minor gases, is involved in a complex global cycle. It is released from the interior of the Earth via volcanic eruptions, and by respiration, soil processes, combustion of carbon compounds and oceanic evaporation. Conversely, it is dissolved in the oceans and consumed during plant photosynthesis. Currently, there are 359 parts per million by volume (ppmv) of CO_2 in the atmosphere, a concentration which is continuing to rise due to anthropogenic (man-made) emissions from the burning of fossil fuels and forests.

Methane

Methane (CH_4) is another greenhouse gas, and is produced primarily by anaerobic (oxygen-deficient) processes such as the cultivation of rice paddies or animal digestion. It is destroyed in the lower atmosphere (troposphere) by reactions with free hydroxyl radicals (OH):

$$CH_4 + OH \rightarrow CH_3 + H_2O$$

Like CO_2, its concentration in the atmosphere is increasing due to anthropogenic activities such as agricultural practices and landfills.

Nitrous Oxide

Nitrous oxide (N_2O) is produced by both biological mechanisms in the oceans and soils, and by anthropogenic means including industrial combustion, vehicle exhausts, biomass burning and the use of chemical fertilizers. It is destroyed by photochemical reactions (involving sunlight) in the upper atmosphere (stratosphere).

Ozone

Ozone (O_3) in the stratosphere filters out harmful ultra-violet radiation from the Sun, and so protects life on Earth. Recently, there have been fears about the destruction of the ozone layer, principally over the Antarctic but increasingly now over the Arctic regions. The concentration of O_3 in the atmosphere is not uniform, unlike other trace gases, but varies according to altitude. O_3 is formed during a photochemical reaction involving solar ultra-violet radiation, an oxygen molecule and an oxygen atom,

$$O_2 + O + M \rightarrow O_3 + M$$

where M represents the energy and momentum balance provided by collision with a third atom or molecule, for example oxides of nitrogen (NOx). The destruction of O_3 involves the recombination with atomic oxygen, via the catalytic effect of agents such as OH radicals, NOx and chlorine (Cl, ClO) radicals. The concentration of O_3 is determined by the finely balanced equilibrium of formation and natural destruction.

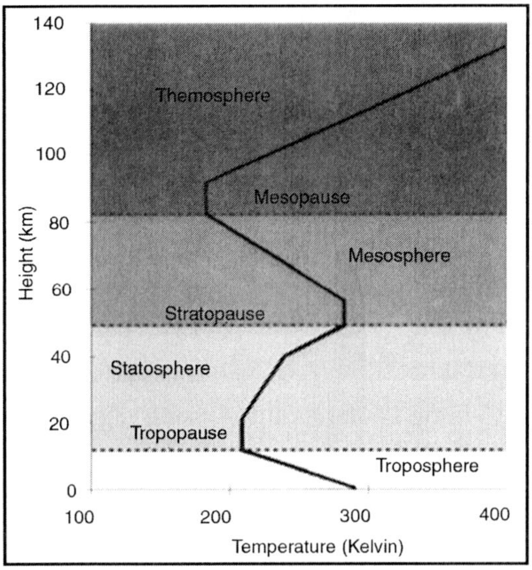

Because the relative reaction rates of formation and destruction vary with temperature and pressure, so O_3concentrations vary with altitude. Most of the ozone occurs in a layer between 15 to 35 km altitude where the relative reaction rates of formation and destruction rates are most conducive to O_3 formation. The present fear about ozone depletion is due to the increase in amount of agents (such as Cl) in the atmosphere that increase the destruction rate of O_3, so upsetting the delicate equilibrium that exists.

Halocarbons

Halocarbons are compounds containing carbon, halogens such as chlorine, bromine and fluorine, and sometimes hydrogen. They may be wholly anthropogenic such as the CFCs or they may have natural sources, such as some of the methylhalides.

Chlorofluorocarbons (CFCs) are entirely anthropogenically produced by aerosol propellants, refrigerator coolants and air conditioners. They are made up of carbon, chlorine and fluorine molecules. CFCs are destroyed slowly by photochemical reactions in the upper atmosphere (stratosphere). CFCs were absent from the atmosphere before the 1930s, but over the last half century, their concentrations have steadily increased.

Although their concentrations are measured in parts per trillion (by volume), they are seen as a significant threat to future global warming. They possess long atmospheric lifetimes measured in decades to centuries, and molecule for molecule, are thousands of time stronger as a greenhouse gas than CO_2. Halons are similar anthropogenic species but contain bromine instead of chlorine.

Other halocarbons include the Hydrochlorofluorocarbons (HCFCs) and hydrofluorocarbons (HFCs). These are anthropogenic compounds currently being used as substitutes to replace CFCs which are being phased out under the terms of the Montreal Protocol, to protect the ozone layer. Many of the other gases in the atmosphere are the noble orinert gases, including argon, neon, helium, krypton and xenon. Because of their chemical inertness, they play little or no part in the chemical and physical processes operating in the atmosphere.

Trace Gases

In addition to these gases, water vapour (H_2O) is a vital atmospheric constituent, averaging about 1% by volume, with significant variations across spatial and temporal scales. Its presence in the atmosphere forms part of the global hydrological cycle. Water vapour, being the most important natural greenhouse gas on account of its abundance, plays a crucial role in the regulation of the atmosphere's energy budget. Despite this, the total volume of water in the atmosphere is relatively small, and, if precipitated completely and evenly over the whole Earth, would yield only about 25mm or 1 inch of rainfall. In reality, of course, rainfall distribution is highly uneven, due to the internal dynamic processes within the global climate system. In addition to the gases there are other reactive gas species produced by cycles of sulphur (S), nitrogen (N_2) and chlorine (Cl) halogens.

Aerosols

Variations in the abundance of the atmospheric greenhouse gases have the potential for changing the global climate. Variations in another group of species, namely the atmospheric aerosols, can also affect climate. Aerosols are solid or liquid particles dispersed in the air, and include dust, soot, sea salt crystals, spores, bacteria, viruses and a plethora of other microscopic particles. Collectively, they are often regarded as air pollution, but many of the aerosols have a natural origin. Although atmospheric turbidity (the abundance of aerosols) varies over short time scales, for example after a volcanic eruption the long term it maintains a fair degree of equilibrium, owing to the natural cleansing mechanisms of the Earth's climate system. Cleansing is never complete, however, and there always remains a background level of atmospheric aerosols which reflects the dynamic processes involved with aerosol input and aerosol removal. Natural sources of aerosols are probably 4 to 5 times larger than anthropogenic ones on a global scale, but regional variations of anthropogenic emissions may change this ratio significantly in certain areas, particularly in the industrialised Northern Hemisphere.

The Vertical Structure of the Atmosphere

Most of the gaseous constituents are well mixed throughout the atmosphere. However, the atmosphere itself is not physically uniform but has significant variations in temperature and pressure with altitude. shows the structure of the atmosphere, in which a series of layers is defined by reversals of temperature. The lowest layer, often referred to as the lower atmosphere, is called the troposphere. It ranges in thickness from 8km at the poles to 16km over the equator, mainly as the result of the different energy budgets at these locations. Although variations do occur, the average decline in temperature with altitude (known as the lapse rate) is approximately 6.5°C per kilometre. The troposphere contains up to 75% of the gaseous mass of the atmosphere, as well as nearly all of the water vapour and aerosols, whilst 99% of the mass of the atmosphere lies within the lowest 30km. Owing to the temperature structure of the troposphere, it is in this region of the atmosphere where most of the world's weather systems develop. These are partly driven by convective processes that are established as warm surface air (heated by the Earth's surface) expands and rises before it is cooled at higher levels in the troposphere.

The tropopause marks the upper limit of the troposphere, above which temperatures remain constant before starting to rise again above about 20km. This temperature inversion prevents further convection of air, thus confining most of the world's weather to the troposphere.

The layer above the tropopause in which temperatures start to rise is known as the stratosphere. Throughout this layer, temperatures continue to rise to about an altitude of 50km, where the rarefied air may attain temperatures close to 0°C. This rise in temperature is caused by the absorption of solar ultraviolet radiation by the ozone layer. Such a temperature profile creates very stable conditions, and the stratosphere lacks the turbulence that is so prevalent in the troposphere.

The stratosphere is capped by the stratopause, another temperature inversion occurring at about 50km. Above this lies the mesosphere up to about 80km through which temperatures fall again to almost -100°C. Above 80km temperatures rise continually (the thermosphere) to well beyond 1000°C, although owing to the highly rarefied nature of the atmosphere at these heights, such values are not comparable to those of the troposphere or stratosphere.

20TH CENTURY CLIMATE CHANGE

There is now considerable evidence that indicates a relationship between the man-made enhancement of the Earth's natural greenhouse effect through greenhouse gas pollution of the atmosphere and the global warming that has been observed.

Measurements of surface temperature recorded around the world during the last 150 years indicate that global temperatures are now higher than in any decade over this period. During the 20th century a global average surface temperature increase of about 0.6°C years has taken place, although the warming trend has not been smooth and has taken place rather differently between the Northern and Southern Hemispheres. With this in mind, it is important to recognise that global average surface temperature, as a measure of the global climate, represents an over-simplification. Winter temperatures and night-time minimums for example, may have risen more than summer temperatures and daytime maximums.

With higher global temperatures one would expect an increase in rainfall and other forms of precipitation, because of the greater amount of moisture available within the atmosphere. Striking changes in precipitation have occurred on regional scales, most notably the drought in the African Sahel between the 1960s and 1980s. Nevertheless, the accuracy of many precipitation records should be treated with caution. Precipitation is more difficult to monitor than temperature due to its greater geographical variability.

Variations in land- and sea-ice coverage and the melting or growth of glaciers occur in response to changes in temperature, sunshine, precipitation, and for sea-ice changes in wind. Since 1966 Northern Hemisphere snow cover maps have been produced by the United States using satellite imagery. Consistent with the surface and tropospheric temperature measurements is the decrease (by about 10%) in snow cover and extent since the late 1960s. There has been a widespread retreat of mountain glaciers in non-polar regions during the 20th century. Variations in sea-ice extent have also been reported, with spring and summer sea-ice extent in the Northern Hemisphere decreasing by between 10 an 15% since the 1950s. Considerable interest is now focusing on Antarctica, where regional warming, as predicted by climate models, has been more rapid than global warming as a whole. In recent years the summertime disintegration of the Larsen Sea Ice Shelf adjacent the Antarctic continent has been occurring on an unprecedented scale. In view of the rapidity at which it is taking place, such an event has been viewed as a possible signal of global warming.

Increased global cloudiness, as for increased global evaporation and precipitation, would be an expected consequence of higher global temperatures.

It is likely that there has been a 2% increase in cloud cover over mid - to high latitude land areas during the 20th century. In most areas the trends relate well to the observed decrease in daily temperature range (since cloudier nights tend to be warmer and cloudier days cooler).

21ST CENTURY CLIMATE CHANGE

Prediction of climate change over the next 100 to 150 years is based solely on climate model simulations run on computers. The vast majority of modelling has concentrated on the effects of continued man-made pollution of the atmosphere by greenhouse gases, and to a lesser extent, atmospheric aerosols. The main concern at present is to determine how much the Earth will warm in the near future.

Significant results from some of the best climate models available indicate that a global average warming of 0.3°C per decade can be expected to occur during the 21st century, assuming that mankind fails to control current emissions of greenhouse gases, although it could be as high as 0.6°C. In addition regional variations in the patterns of temperature and precipitation change will occur, with greater warming likely in the polar regions. Currently, models suggest that if the atmospheric concentration of carbon dioxide, the main greenhouse gas, is doubled from pre-industrial levels, the Earth will warm by between 1.5 and 4.5°C sometime over the next 200 years or so. The large margin of error in future prediction of temperature emphasises that modelling the climate is inherently a difficult business. Part of the problem stems from trying to guess what climate feedbacks might occur that may enhance the initial warming due to an enhanced greenhouse effect. Melting ice in the polar regions for example could accelerate warming because exposed ground absorbs more energy from the Sun than snow and ice, which reflect about 80 to 90%.

Whilst uncertainties concerning the actual response of the global climate to man-made greenhouse gas emissions exist, most scientists agree that the global warming trend of the 20th century will continue into the 21st century. The projected rate of warming is faster than at any time during recent Earth history. If nations fail to respond, the world may experience numerous adverse impacts as a result of global warming in the decades ahead.

Aerosols

Atmospheric aerosols are very fine particles suspended in air. They are formed by the dispersal of material at the Earth's surface (primary aerosols), or by reaction of gases in the atmosphere (secondary aerosols). They include sulphates and nitrates from the oxidation respectively of sulphur dioxide and nitric oxide during the burning of fossil fuels, organic materials from the oxidation of volatile organic compounds (VOCs), soot from fires, and mineral dust from wind-blown processes. Natural aerosols, which also include sea salt and volcanic dust, are probably 4 to 5 times larger than man-made ones on a global scale, but regional variations in man-made pollution may change this ratio significantly in certain areas, particularly in the industrialised Northern Hemisphere. Although making up only 1 part in a billion of the mass of the atmosphere, they have the potential to significantly influence the amount of sunlight reaching the Earth's surface, and therefore climate. Removal of most aerosols is mainly achieved by rainfall (wet deposition) and by direct uptake at the surface (dry deposition). Explosive volcanic eruptions however, can inject large quantities of dust and gaseous material, such as sulphur dioxide, high into the atmosphere (the stratosphere). Here, sulphur dioxide is rapidly converted into sulphuric acid aerosols. Whereas pollution of the lower atmosphere is removed within days by the effects of rainfall and gravity, stratospheric pollution may remain there for several years, gradually spreading to cover much of the globe.

Like greenhouse gases, aerosols influence the climate. Atmospheric aerosols influence the transfer of energy in the atmosphere in two ways: directly through the scattering of sunlight; and indirectly through modifying the optical properties and lifetimes of clouds. The scattering of sunlight by aerosols is clearly demonstrated in the aftermath of a major volcanic eruption, when exceptionally colourful sunsets may be witnessed. The volcanic pollution results in a substantial reduction in the direct solar beam, largely through scattering by the highly reflective sulphuric acid aerosols. Overall, there is a net reduction of 5 to 10% in energy received at the Earth's surface. An individual eruption may cause a global cooling of up to 0.3°C, with the effects lasting 1 to 2 years.

Estimation of the impact aerosols have on longer-term global climate change however, is more complex and hence more uncertain than that due to the well-mixed greenhouse gases. This is largely because the geographical distribution of aerosols is highly variable and strongly related to their sources. The best estimates of global cooling attributable to man-made aerosols are based on computer models. These show that the global cooling effect of man-made aerosols could offset the warming effect of increased greenhouse gas concentrations by as much as 30%. The variable distribution of aerosols however, makes calculation of a global average difficult. Nevertheless, it is likely that aerosols may slow the rate of projected global warming during the 21st century.

2

Migration and Climate Change

The main purpose of this chapter is to review briefly existing research on the likely impact of environmental change on the movement of people. The topic of environmental migration is not a new topic of research. As early as 1990, the Intergovernmental Panel on Climate Change (IPCC, 1990: 20) warned that the greatest single impact of climate change could be on human migration – with millions of people displaced by shoreline erosion, coastal flooding and severe drought. In 1992 the International Organization for Migration (IOM) published a report on "Migration and Environment" in which it is stated: "Large numbers of people are moving as a result of environmental degradation that has increased dramatically in recent years. The number of such migrants could rise substantially as larger areas of the earth become uninhabitable as a result of climate change" (IOM, 1992).

However, until two or three years ago, the topic of migration and the environment was largely ignored by migration experts and policymakers. Indeed, in the 2005 report of the Global Commission on International Migration, there is barely a mention of the topic.

Part of this neglect may be due to the marginal consensus over the years among researchers about whether or not environmental migration is a distinct form of migration worthy of special study. There has been considerable disagreement about how to conceptualize the relationship between migration and climate change and about research methodologies to be used to investigate the topic further. While it is recognized that there is a two-way relationship between migration and the environment, the main emphasis in this review is on research on the implications of environmental change for migration, rather than vice versa. IOM titled *Migration, Environment and Climate Change: Assessing the Evidence* (IOM 2009). Given limitations of space, the paper focuses on four main research challenges which are discussed in the IOM book:

1. How has the relationship between migration and the environment been conceptualized?

2. To what extent has it been possible to measure the scale of environmental migration?
3. What evidence is available regarding the impact of environmental migration?
4. What research has been conducted on policy responses?

The paper is not limited to a focus on climate change but looks more broadly at the range of environmental factors which impact on migration. This is because many changes in the environment, such as earthquakes, which are not necessarily linked to climate change, have an enormous impact on the movement of people.

MIGRATION IN INDIA

India as a nation has seen a high migration rate in recent years. Over 98 million people migrated from one place to another in 1990s, the highest for any decade since independence according to the 2001 census details. However in 1970s migration was slowing down. The number of migrants during 1991-2001 increased by about 22% over the previous decade an increase since 1951. Apart from women migrating due to marriage, employment is the biggest reason for migration. The number of job seekers among all migrants has increased by 45% over the previous decade. Nearly 14 million people migrated from their place of birth in search of jobs. The overwhelming majority of these-12 million was men.

Migrants have created pressure on others who are in same job market. While freedom to migrate within the country is an enshrined right the uneven development, levels of desperation and other factors have created friction points. Most people migrate because of a combination of push and pull factors. Lack of rural employment, fragmentation of land holdings and declining public investment in agriculture create a crisis for rural Indians. Urban areas and some rural areas with industrial development or high agricultural production offer better prospects for jobs or self-employment.

Contrary to common perception the search for jobs is more often within the same state than in some other state. About 9 million persons were intra-state migrants often within the district while 5 million went to other states. The intra-state figures include people moving from villages to nearby towns and cities in search of better jobs. Over 5.7 million persons who moved in search of jobs migrated from rural to urban areas. Another 4.5 million migrated within the rural areas looking for work.

The data shows that among people migrating in search of jobs, literates constitute the vast bulk over 10.6 million while illiterate migrants are about 3.3 million. Three out of four job-seeking migrants are educated males. Among literate, migrant job-seekers less than 1% was women. Nearly 40% of literate persons migrating for work had studied up to secondary level and another 32% had studied beyond. Graduates numbered over 1.8 million or about 17% while technical diploma or degree holders constituted about 8%.

About 72% do get regular work but over 11 million get less than 183 days of work in a year. This is a higher proportion than non-migrants. Independent NSS data from 1999-2000 indicates that migrant workers take up regular or casual employment or self-employment in nearly equal proportions. Around 8.1 million of the migrants were reported as available for or seeking work. The census data may not fully reflect seasonal or circulatory migration, estimated to be up to 10 million by the National Commission on Rural Labor. Seasonal migrants are usually dalits and other highly impoverished sections that go out to work in harvesting seasons or on construction sites, in brick kilns, salt mines etc. They go out to pay their debts and to survive.

MIGRATION, GOVERNANCE AND SOCIAL CHANGE

In a series of articles published in *Economic & Political Weekly* since 1973, we reported on economic and social change in several villages in Purnia district of Bihar. The last of these articles highlighted some striking developments in production, incomes and labour markets as the old semi-feudal relations gave ground to market forces. In 2008 and 2009, we again visited two of these villages, in Kasba block, to see whether there had been further developments: had the new economic and social relationships which emerged in the 1990s resulted in sustained growth and rising incomes? Had the acceleration of growth in the Indian economy as a whole had visible effects on the village economy? What was the impact of the political and governance changes of the last decade? Bihar, despite recent indications of increased growth, has lagged behind as other parts of India have developed. By 2004-05 output per capita amounted to only 35% of the all-India average and the incidence of poverty was high (55.7% of the Bihar rural population was below the official poverty line against 41.8% of the all-India rural population; GoI 2009). Purnia district is, in turn, relatively backward compared with the State as a whole, with per capita output 25% lower than the Bihar average.

The two villages studied are Pokharia, a small, backward caste-dominated village, with 918 inhabitants in 2001; and Dubaili Biswaspur, a larger village with 4,689 inhabitants in 2001. In Dubaili Biswaspur we concentrated mainly on one *tola*, Mazgama West, accounting for about a quarter of the households in the village, with a backward class Muslim majority and a smaller number of scheduled caste (SC) households. We make no claim that these villages are representative of Bihar or even of their district. But neither is there an obvious bias in the choice of these villages. The patterns which we observe are reproduced elsewhere in the region, and our findings can help identify issues and impacts which merit closer attention more generally.

In the 1970s and early 1980s, these villages were backward and stagnant, and poverty was intense. Wages barely sufficed to cover basic subsistence, and real incomes were if anything declining. Mortality was high and production relations "semi-feudal", in the sense that debt bondage, tenancy and attached labour were widespread, served as mechanisms of labour control and exploitation, and were resistant to change. Communications were poor, facilities limited, education levels low. The government action was extremely weak.

From the mid-1980s, things began to change, and by 1999, the picture had altered substantially. Short-term migration to northwestern India seemed to be an important trigger of change, undermining feudal relationships, creating new perspectives, generating additional income sources and pushing up local wages. Diets had improved, and to some extent, housing. Infrastructure had also improved – roads, irrigation, schools – and there had been a significant increase in agricultural productivity. At the same time, the transition was clearly incomplete, with various remnants of the semi-feudal labour and reward systems still present. Agricultural innovation was limited, gender inequality strong, and there was not much visible impact of government programmes.

This paper explores the impact of another decade of development. After a brief introduction to the villages it examines changes in agrarian relations, looks in more detail at migration and its impact, and considers trends in governance and in the effectiveness of state policies.

THE VILLAGE ENVIRONMENT

Population densities are high in Bihar, and these villages are no exception – according to the census data for 2001, 695 persons per sq km in Pokharia and 1,379 in Dubaili Biswaspur (compared with 881 for Bihar as a whole and 312 for India). Moreover, population growth is rapid. Between 1981 and 2001 the population of Pokharia almost doubled and that of Dubaili Biswaspur increased by over 75%. Literacy rates have been rising but at a very slow pace.

According to the 2001 Census, even after excluding the 0-6 age group (illiterate by definition), over 50% of men and over 80% of women were illiterate in both villages. The last decade has, however, seen a revolution in school enrolment. New schools have opened and school attendance, especially of girls, has risen rapidly.

Within the villages, occupation remains dominated by agriculture, and the proportion of agricultural labourers has been rising – from 31% of all (main) workers in 1981 to 60% in 2001 in Pokharia, and from 50% in 1981 to 72% in 2001 in Dubaili Biswaspur according to the census.

In 2001, 28% of workers were classified as cultivators in Pokharia and 22% in Dubaili Biswaspur. In Pokharia holdings are small – no household had more than five acres in 2009. In Dubaili Biswaspur some larger holdings were reported, but only six to eight households out of the 215 in Mazgama West tola were reported as having more than eight acres. Apart from traditional occupations such as basket-making, fishing and other caste occupations, there was little non-agricultural employment in these villages.

There are few *pucca* houses, and most houses are made of traditional materials, mud and thatch (though, often now, with a corrugated iron roof), while many of the poorer households still live in simple constructions of bamboo and thatch.

Access to drinking water has improved since 1999 with an increase in the number of handpumps, though some problems of water quality subsist, especially in Pokharia, where one-third of households do not have their own supply, and in the SC locality of Mazgama.

Village roads and paths and drainage show little sign of improvement, and sanitation facilities are still virtually non-existent. The most striking and visible change concerns communications. Hard to reach in the 1980s, today the villages are only 6 km, along a good pucca road, from a four-lane dual carriageway.

This means that they are within easy reach of the local block headquarters throughout the year, and also only 40 minutes by autorickshaw from the main district market which is in turn only a few kilometres from the district headquarters.

There were many more electricity connections in 2009 than in 1999, both legal and illegal, although electricity is available only four to five hours per day and is not used for agriculture because of its unreliability. A few solar street lights have been installed, mainly outside the houses of influential villagers. There is a mobile phone in the majority of households, even among the poorest groups. There has been some growth in the number of shops, with seven in Pokharia and three in Mazgama and several more on the road nearby, in addition to a weekly market.

CHANGES IN AGRARIAN RELATIONS

The key change observed in 1999 was the shift towards market relationships, especially in the labour market. Since earlier research had connected agricultural backwardness with semi-feudal social and economic relationships, there was reason to expect that market forces, combined with improved communications, would unleash the unfulfilled agricultural possibilities of a region with abundant groundwater and reasonable soils, and a considerable potential for triple cropping.

CONCEPTUALIZING THE RELATIONSHIP BETWEEN MIGRATION AND THE ENVIRONMENT

The migration and environment research literature tends to fall into two broad and extreme categories:

1. work done by "minimalists" who suggest that the environment is only a contextual factor in migration decisions and
2. work done by "maximalists" who claim that the environment causes people to be forced to leave their homes (Fraser et. al., 2008).

Although many experts accept that climate change is a factor which can impact the decision to migrate, the conceptualization of this factor as a primary cause of forced displacement has been questioned (Black, 2001). While the environment can be a driver of migration, more often than not a complex combination of causes determines whether or not people move. Given the multiple causes of migration, drawing a clear line between voluntary and forced movements is not always straightforward.

This disagreement on the role of the environment in inducing migration is reflected in further disagreement over terminology. It is common to describe those who move for environmental reasons as climate change refugees or as environmentally displaced persons and to characterize such movements as forced migration. Popular with the media, the term "environmental refugees" has been used to describe the whole category of people who migrate because of environmental factors. This broad definition, while evoking an image that has brought public attention to the issue, is not sufficiently precise to describe all the various types of movements which may be linked to environmental factors. In some situations, such as natural disasters, people may have little choice but to move, and may be forcibly displaced. In other situations where environmental change is gradual, movement is more likely to be voluntary as people have time to weigh their options, and environmental change may be one of many factors inducing them to move.

It is perhaps more useful, instead, to think in terms of a continuum:

"Population mobility is probably best viewed as being arranged along a continuum ranging from totally voluntary migration... to totally forced migration, very few decisions are entirely forced or voluntary" (Hugo, 1996).

In the absence of an internationally agreed definition, IOM developed a working definition in 2007 which defines "environmental migration" as follows: "Environmental migrants are persons or groups of persons who, for reasons of sudden or progressive change in the environment that adversely affects their lives or living conditions, are obliged to leave their habitual homes, or choose to do so, either temporarily or permanently, and who move either within their country or abroad."

The purpose of this definition is to try to encompass population movement or displacement, whether it be temporary or permanent, internal or cross border, and regardless of whether it is voluntary or forced, or due to sudden or gradual environmental change.

Measuring the Scale of Environmental Migration

How many people have been migrating in recent years due to environmental change and how many more people are likely to migrate in the future? There are no firm answers to these questions, but it is assumed that most of the migration that will occur will be mainly within those developing countries which are likely to be most affected by climate change. Today, approximately three-quarters of all migrants move within borders, and only 37 per cent of migration in the world is from developing to developed countries (UNDP, 2009).

Probably, the best available data on environmental migration are data on the numbers of persons displaced as a result of natural disasters. In 2008, for example, it has been calculated that 20 million people were displaced by sudden-onset climate-related extreme weather events (OCHA-IDMC, 2009). However, even in the case of natural disasters where better data exists, we have no global data on migratory movements related to natural disasters (Hugo, 2008). At best, there are estimates that can be derived from displacement data relating to particular crises (Naik, 2009). Despite the fact that the reported number of disasters has doubled over the last two decades (Basher, 2008) we have not seen a major impact on international migration flows, as much displacement is short distance and temporary. The Tsunami which hit Asia in 2004 claimed the lives of an estimated 200,000 persons and displaced around 400,000 others, yet the vast majority of those who were forced to move relocated to nearby areas.

Extreme environmental events such as cyclones, hurricanes, earthquakes, tsunamis and tornadoes tend to capture the media headlines, but gradual changes in the environment may have a much greater impact on the movement of people in the future. For example, during the period 1979 to 2008, 718 million people were affected by storms compared to 1.6 billion people affected by droughts (EM DAT, 2009). Unfortunately, however, there is relatively little information on the links between numbers affected by gradual changes in the environment and migration.

Reasons for Lack of Statistics

There are several factors which make it difficult to predict the likely scale of environmental migration (Brown, 2008). First, it is difficult, as mentioned earlier, to disaggregate the role of climate change from other economic, political and social factors which drive migration. Second, there is a basic lack of migration data available in developing countries which are likely to be most vulnerable to climate change. For example, in a recent report, the Commission on International Migration Data for Development Research and Policy noted that many countries still do not include basic questions about migration in their censuses. Even in the ongoing 2010 census, several countries including Japan, Mexico, Korea, the Philippines, and Egypt, do not include questions on the place of birth. One-third of countries also do not ask about previous residence in another country (CGD, 2009).

Third, the lack of data is largely due to the absence of an adequate definition to cover migrants affected by natural disasters under international law.

Fourth, it is extremely difficult to predict the impact of climate change. Climate modeling techniques have not yet even begun to account adequately for the impact of individual choice, the potential for international action and the variability of future emissions and meteorological scenarios (Brown, 2008).

Evidence of the Impact of Environmental Migration

There is relatively little reliable data on the impact of environmental migration because there has been relatively little empirical research on this topic. There are many good studies on the impact of migrants on environments, on land use, deforestation, and so forth, but there is almost no reliable evidence on the effects of environmental factors, "controlling for other influences, on out-migration, particularly from rural areas" (Bilsborrow, 2009). In a review of literature over the past 50 years, of 321 publications, including 153 articles in peer-reviewed journals and 29 books, only two articles were found which investigate the effects of environmental factors on out-migration based on quantitative multivariate methods (Moriniere, 2009).

Relatively few social scientists who focus on migration – and rely on data from censuses and household surveys – have been engaged in data collection or research on the environment (IOM, 2009). Moreover, the little research that has been conducted on the effects of environmental change on migration has tended to focus on the negative consequences of environmental migration. Few studies have explored how migration can be a coping or adaptation strategy or how migration can relieve pressure on environmentally degraded areas (IOM, 2009), as suggested in studies in countries such as El Salvador,

Jamaica and the Philippines that have found that migrants respond to extreme environmental events by increasing their remittances (UNDP, 2009).

Research on Policy Responses

Research on policy responses to environmental migration is in its infancy. There has been little analysis of what "standards, policies or programmes are most appropriate for managing this category of internal or international migration flows" (Leighton, 2009). As most environmental migration is expected to occur within and between developing countries in the South, there has been little incentive for policymakers in destination countries in the North to adjust their immigration policies. Few destination countries have elaborated specific policy measures to respond to environmental migration, and none currently have a pro-active policy to resettle those affected by environmental disasters (Martin, 2009). At best, policies are ad hoc, with some countries taking measures to allow migrants to remain temporarily in the destination country when disasters occur at home.

A recent review of current policy responses in destination and origin countries finds that there is also little coherence between environmental change and migration policies (Martin, 2009). For example, few of the major middle-income developing countries which are major source countries for migrants, such as Mexico, India and China, have included any reference to migration in their climate change adaptation plans. Although many countries clearly lack adequate resources to respond to the growing number of natural disasters, the international community has at least established a policy framework for responding to emergencies. On the other hand, a strategy and policy framework to address the impact of gradual environmental change is largely lacking, and would require linking development, environment and migration policies in a much more coherent manner.

Analysis of policy responses to environmental migration has also tended to focus much more on responses to extreme environmental events rather than on how best to manage the impact of gradual changes in the environment on population mobility. Research tends to focus on questions such as how best to provide emergency assistance to those who are displaced, how to reduce disaster risk and how to improve the legal and normative framework for the protection of the displaced. A number of studies have also discussed whether there is a case for introducing a new set of legal instruments to protect the environmentally displaced. Zetter (2009), in a review of this discussion, concludes that there is little opportunity or need to create an entirely new set of legal instruments to address environmental migration. There is much scope within existing legal frameworks to provide protection to those who are forced to move for environmental reasons, but there is a critical lack of capacity in many states to implement existing frameworks and it is here where there is a need for much more research.

POPULATION, MIGRATION AND DEVELOPMENT

The relationship between MDGs and population trends and structures has started to be considered, but there is still need to improve our knowledge of how demographic mechanisms, including migration, influence progress towards MDGs attainment.

This book will mostly consider the situation in Pacific island countries and major migration countries in South, East and South-East Asia, including China and India. While the relationship between population, migration and MDGs is difficult to assess in large and interdependent Asian countries, linkages appear more clearly in small Pacific countries where the impact of migration is much more important. Although lessons learnt form the Pacific may not be replicable in larger economies, it contributes to improving our knowledge of the effects of migration on population and MDGs.

MIGRATION TYPES

Migration types are not neutral on its impact on population and development. In Asia, due to lack of political agreement on long-term and permanent migration schemes, such as green cards, migration consists mostly of 'guest workers' on short-term contracts that can be extended (Asis 2005, Abella 2005). Singapore has developed medium and long-term contracts for highly qualified migrants. Chain migration patterns have developed with unskilled construction workers migrating from Myanmar to Thailand; semi-skilled Thai workers migrating to Taiwan, Singapore, while Malaysians migrate to Singapore, Korea and Japan. Family reunification exists mostly for high qualified migrants. However, schooling of migrants' children has been addressed by several host countries including Singapore, Japan and Korea. Student and work contract migrations have consequences on age and sex of migrants. Students are youth; males migrate more often for industry jobs and women for services jobs. Age plays different roles according to types of migration as rim countries migration schemes grant points according to age. Despite forecasts of increasing shortages on labour markets in host countries (Mason), migration policies are still mostly oriented toward a restrictive selection process and temporary migration. This certainly reduces the level of legal flows and result in higher undocumented migration the characteristics of which are less precisely known. In the Pacific, some Polynesian countries, except Tonga, are former New Zealand colonies and benefit from special access to New Zealand: Cook Islanders, Niueans, Tokelauans are all New Zealand citizens, and Samoans benefit from a special quota. Tuvaluans and Tongans (in Polynesia) and Kiribati (in Micronesia) also have some access to New Zealand under the Pacific Access Category (PAC), as does Vanuatu (in Melanesia)i (Bedford and al. 2006, Bedford 2005). Former US affiliated islands: FSM (Federated States of Micronesia) and RMI (Republic of the Marshall Islands) have special access to the US and to US territories (Guam and American Samoa). The special schemes have resulted in large scale migration, with free movements to New Zealand for selected groups of Polynesians, extended to Australia for New Zealand residents in the frame of the Trans-Tasman Travel Agreement, and resulting in the development of transnational communities with frequent circulation, as well as overstaying, between islands and New Zealand, further including Australia and the USii. Large scale circulation also applies to Micronesian migration towards the US. These two networks are typical of Pacific migration. Such schemes are made possible by the small size of island states and have resulted in mass migration from some island countries. But, although free movement would enable return migration and circulation, this has little developed beyond visits for Christmas and family events, and it has not resulted in much brain circulation that would enable more rapid economic development.

Labour migration has a long history in parts of the Pacific, initially for work on plantations, later for work in phosphate mines, and more recently through recruitments of seamen from Tuvalu, Kiribati and to a lesser extent, Fiji. There is also migration from Fiji to UK to work in military and to the Middle East to work as security personnel. Fiji is also largely affected by brain drain, with nurses migrating to rim countries and secondarily to Middle East. Shortages in nurses in Fiji have resulted in migration of Pilipino nurses. While Cook Islanders migrate to New Zealand, shortages in the tourism industry are filled by Fijians on short-term contracts. These are Pacific cases of chain migration.

DEMOGRAPHIC IMPACT OF MIGRATION

The impact of migration on population trends and structure is well known but flows have drawn most of the attention and, actually, a precise measure of flows is needed to assess the demographic impact of migration. Beyond uncertainties on the size of flows, data on stocks are also unavailable, with Singapore and Malaysia, among others, not publishing data on nonresident population. Would such data be available, comprehensive estimate of the impact of migration in a long-term perspective - answering questions like: what would be the population of the Philippines or the Cook Islands if there had been no migration since 1960 or 1970 - would be purely theoretical because fertility would certainly have been different on the long-term had migration not taken place. More simply, we shall try to consider the current impact of migration on population trends and structure from the latest censuses in Asia and the Pacific, with more attention to a few countries. Consequences of migration on population growth are well known. They include, as regards purely demographic impacts:

- Reduced population growth for emigration countries and higher growth for immigration countries - the former is usually considered favorable to developing countries;
- Changes in sex ratios according to gender differentials in migration patterns, with various directions being possible in both emigration and immigration countries - imbalances are usually considered unfavorable;
- Changes in age structures: increased dependency in emigration countries and reduced dependency in immigration countries – the former is considered unfavorable in the frame of the demographic window theory (Mason 2006, Mason 2001, Bloom, Canning 2001). Social and economic impacts are many, but the most commonly stated relate to:
- The labour force: depleted working ages and brain-drain from emigration countries and increased labour force and brain-gain in immigration countries, the former are considered unfavorable - however, migration releases tensions on the labour market;
- Family breakdowns reflected by increased headship rates for females in emigration countries and more frequent lone persons and 'not related' or 'other relatives' members of households in host countries – the direction of related effect is not well established

The impact of migration at regional level is more rarely mentioned but it deserves to be considered, mostly as it involves populations of very different size, like in the case of China and India on one side and South and South-East Asia on the other. Economies have become interdependent in the frame of a well established system of regional migration. In emigration countries with labour surplus, migration is a safety valve that reduces unemployment and underemployment, while in immigration countries with labour shortages, migration supports economic growth – however labour surpluses and shortages need to be considered at sectoral level. The beneficial effect of emigration as a safety valve is limited in larger countries like China and India that can 'export' only a minor part of their unemployment and under-employment because receiving countries have not the capacity to absorb it and need to protect their labour market through migration and work policies (Rallu, 2001). For instance, in China, rural underemployment was estimated to be well above 100 millions people and the region, not even the world economies cannot integrate such numbers of unqualified migrants. Nevertheless, the migration pressure of China in the region remains a concern for smaller South-East Asian countries, with Chinese setting up retail businesses and taking a large share of small trade. While there is a regional circulation of labour force at various levels of qualification and sectors of the labour market, there is also an extra-regional migration to Middle East or to developed countries of the Pacific rim and the West. Thailand, Myanmar, Malaysia, Indonesia, Bangladesh, Pakistan and Sri-Lanka are largely involved in intra-regional migration (as well as extra regional migration to the Middle-East). But China, India, the Philippines and Viet-Nam have a large part of their migration directed towards rim countries and to the West.

At intra-Pacific level, migration of Pacific islanders is quite limited with only Fiji, Cook Islands and Palau attracting immigrants: mostly through regional organizations and the University of the South Pacific for Fiji and in tourism industry and construction in the Cook Islands (mostly Fijians) and Palau (mostly Micronesians). Fiji and Palau also have Asian migrants, mostly Filipinos, on work contracts in health and tourism. However, Cook Is and Fiji are mostly emigration countries. Palau experienced increasing emigration of its youth and Asian net migration declined in 2000-2005. Most of Pacific islanders' migration is directed towards rim countries: Australia, New- Zealand and the USAiii as well as to US and French territories but the latter rather consists in more or less closed US and French regional networks, that however include Asian migration, mostly for US territories. The mass migration from the Pacific, in the frame of preferential access to rim countries, results in important reduction of population growth in island countries.

Population Growth

The effect of migration on population growth varies greatly by countries, small countries being more affected than large countries.

In Asia, net migration remains below 2 per 1,000 and often below 1 per 1,000 in absolute values in most countries and therefore only slightly impacts on population growth. Its impact on China and India (-0.3 per 1,000 and -0.2 per 1,000) is even much lower (table 2). The highest negative net migration rate is observed in Sri Lanka reaching close to -0.5 per cent and the highest positive net migration is observed in Singapore reaching close to 1 per cent (it was close to 2 per cent for the 1995-2000 period) – Brunei Darussalam has migration rate of 2 per 1,000. In the Pacificiv, net migration nearly erases the effect of natural growth in FSM, Nauru, Samoa, Tonga, or even inverses growth in Niue, Tokelau and occasionally in Cook Isv (Rallu, Ahlburg, forthcoming). In those countries net migration is frequently close or even above 2 per cent. Migration also considerably reduces population growth in RMI but it has declined recently. In Fiji, migration, mostly of Indo-Fijians but also increasingly of indigenous Fijians (with emigration rate of -1.7 per cent the former and -0.2 per cent for the latter; -1.0 percent for Fiji), reduces growth substantially and results in population decline for Indo-Fijians. Seamen migration from Kiribati had not much impact on growth as returns at end of contracts and some returns from Nauru tend to equilibrate flows. Palau had higher net migration rate in 1995-2000 (1.2 per cent) lifting total growth to 2.1 per cent, but it has much reduced in 2000-2005.

Sex Ratios

In the past, mostly males migrated and the analysis of sex ratios imbalances was used to reveal migration. But female participation in migration has been increasing for several decades and now many countries have higher female than male migration. Sex ratios have become difficult to interpret in regards of migration, mostly when emigration and immigration are present together. However, a few countries, like Indonesia, still have predominantly male migration and show low sex ratios at young adult ages. In the Pacific, ancient emigration countries (Tonga and Samoa) have turned to predominantly female youth migration, but mid-adult ages in Tonga show clearly imbalances in sex ratios linked with mostly male migrants in the early stage of the migration process. Predominantly male migration to Palau steeply increases sex ratios, reaching 139 males per 100 females at ages 25-44.

Age Structures and Dependency Ratios

The depletion of young adult ages linked with migration is a major demographic impact of migration. It is associated with change in dependency ratios. However, dependency is also strongly affected by the level of fertility. The impact of migration at young adult ages is sometimes difficult to see on agepyramids, due to past changes in fertility and various events affecting the history of countries, like conflict for Sri-Lanka. When migration affects a large range of ages like in the Philippines, the impact is not much visible. It is much clearer in Indonesia for males and Sri-Lanka for both sexes. The depletion of young adult ages is most typical in mass migration island countries where it results in bottle neck shaped age-pyramids, like in FSM, Cook Islands and Palau. In Palau, both youth emigration of both sexes and predominantly male immigration at mid and late adult ages are well visible.

Cohort Change

A more interesting approach to the impact of migration on population consists in cohort change. While age-pyramid and sex ratios did not show much effect of migration on population structure in the Philippines, changes in cohort size at ages under 50vi clearly show emigration in cohorts reaching ages 15 to 34. The impact was limited at ages 20-24 for males and 20-34 for females in 1995-2000, but it increased and extended to ages 30-34 in 2000-2005 for both sexes. In both periods, net losses are more important for females than for males. Some return migration appears at ages 35-39, but trends thereafter are difficult to interpret as combined effects of immigration/return migration and emigration can occurvii. Results may also be affected by quality of age reporting or age selective variations in coverage of enumeration.

Similar data for Pacific islands show much higher net losses starting as early as late teen ages, with one third of female youth cohorts and close to 40 per cent of males 25-29 emigrating from Samoa (and a similar situation occurs in Tonga and Cook Is), 20 per cent and 25 per cent of respectively female and male Marshallese aged 20-24, and between 15 per cent and 20 per cent of the 20-34 years old leaving Fiji, with higher rate at ages 30-34 in the frame of post-coup emigration that affected many adult Indians. In Palau, while all cohorts showed net migration increases between 1995 and 2000, only the cohorts reaching 25-29 in 2005 show net migration in 2000-2005, with emigration of Palauans at 20-24 and net departures of Asian migrants at ages above 30.

Such migration levels look like an exodus from small island countries and the situation in Sri Lanka and smaller Asian countries is intermediate between those of larger S-E Asian and small islands countries. At high levels of emigration, the impact on population structure and dependency is considerable.

Dependency Ratios

The increase in dependency due to migration is the result of depleted adult cohorts. Their impact on dependency can be high and it directly affects potential for development in the frame of the theory of the demographic window of opportunity. Recent studies have shown that low dependency is favorable to economic growth (Mason 2006, Mason 2001, Bloom, Canning 2001), on condition it is accompanied by high employment level of the youth bulge, which is possible in the frame of friendly economic and investment policies as well as social and political stability and good governance. But, emigration influences dependency ratios in the wrong direction. However, emigration countries have their workers abroad sending remittances that are usually higher than what they would earn in-country.

It would need controversial reconstruction of population trends for decades with and without migration to estimate the long-term impact of migration on dependency ratiosix. Therefore, the impact of migration cannot be estimated comprehensively and precisely, but it is possible to compare dependency ratios according to countries' migration status. In Asia, Singapore, China and Thailand have the lowest dependency and also the lowest fertility among countries listed in table 3. Among countries that have rather similar levels of migration like Bangladesh and Indonesia, the latter has much lower dependency due to completed fertility transition while the former has still TFR above 3. Similar characteristics and gaps prevail between the Philippines and Viet Nam, with the latter having both lower emigration and fertility resulting in lower dependency. And despite higher migration than the Philippines, Sri Lanka has lower dependency due to completed fertility transition. Further comparison and interpretation is difficult as several host countries (Malaysia, Singapore) do not include migrants in their census reports that relates to de-jure population only. Altogether, it appears that dependency is still strongly affected by fertility levels and the related children burden and the role of migration is difficult to measure, although its impact is evident.

In the Pacific, Palau is the only country that has already achieved fertility transition (with TFR of 1.9 in 2001-2005); it also has significant immigration and consequently shows the lowest dependency ratio in the region, similar to China and Thailand. In Fiji, Indians have achieved fertility transition but they have high emigration; their dependency ratio (55 in 1996) was much above that of Palau, but still much lower than for indigenous Fijians (70) who still have TFR of 3.3 and much lesser migration All other countries, except PNG, Solomon Is and Vanuatu are more or less affected by emigration. Samoa and Tonga have TFR just above 4 since the late 1970s and the mid 1980s respectively but they have ancient and important emigration. Their dependency ratio is similar to those of Solomon Is and Vanuatu that have witnessed fertility decline much more recently with TFR still close to 5 at the end of the 20th century. Despite similar, or even more pronounced, migration patterns than Tonga and Samoa, the Cook Islands have lower dependency due to lower fertility. FSM and RMI had both recent fertility declines and important migration; they also show dependency levels close to those of Melanesian countries.

Trends show that stabilization of the size of the population in the most affected migration countries in the Pacific goes hand in hand with nearly stable dependency ratios. Only RMI, FSM and to a lesser extent Fiji and Tokelaux show significant declines in dependency that are linked with fertility declines. Other countries show stable (Tonga), fluctuating (Cook Is, Niue) or slightly increasing (Samoa, Tuvalu) dependency ratios at a rather high levels, mostly when fertility remains high with TFR around 4. However, such stable fertility levels are necessary to avoid population decline. Altogether, these countries have achieved quasi stable populations through both quasi stable fertility and migration. As remittance based economies, achieving the demographic window does not seem to be felt as an issue in those countries. Thus, it appears that migration can erase the benefits of fertility decline for high emigration countries for a long period of time, eventually resulting in quasi stable population size and structure characterized by high dependency in the Pacific. Whenever high fertility is a major factor of high dependency in the Pacific, as well as in Asia, it appears that migration can delay the onset of the demographic window of opportunity, with most origin countries having dependency ratio well above 60 and as high as 80. The impact of migration on labour force and gender empowerment will be addressed in relation to MDGs. The above analysis of the impact of migration on population will help us understand its impact on MDGs attainment.

IMPACT OF MIGRATION ON MDGS

MDGs are development indicators and some of them have been recognized as such for decades, like infant mortality rate (IMR), as well as maternal mortality ratio (MMR) that appears to be closely correlated with the former in both Asia (Choe, Chen 2006) and the Pacific, and both are also correlated with the coverage of services in health. Other MDGs are obviously and closely linked to development, like poverty and environment (access to improved water and sanitation) indicators. Education is also a well acknowledged factor of development as well as of reduction of infant mortality and morbidity, mostly in relation to female education as well as employment. Thus, this section will address the impact of migration on development through its impact on population and MDGs indicators.

Services Indicators in Education, Health and Environment

The first demographic impact of emigration is reduced population growth and it is considered to have favorable effects. The impact of rapid growth on MDGs services indicators: primary and secondary education enrolment, immunization, skilled attended deliveries, is usually negative. In the Pacific, countries that have high population growth rates usually have lower coverage of services than those with low growth and they face the challenge of increasing coverage while births and children cohorts are increasing – not withstanding necessary improvement in the quality of services. This is clearly the situation in Solomon Is, Vanuatu and other rapid growth countries that have no emigration outlet. In migration countries like Samoa and Tonga, despite TFR still around 4, migration of reproductive age adults has resulted in stable birth cohort size since several decades. Thus, it has been possible to increase coverage of services close to 100 per cent and there is no difficulty keeping this level steady. However, this is realized through fragile balance between fertility and migration. Despite emigration, high and slowly declining fertility in the Philippines resulted in increasing birth cohort size until the 1995 census. The 2000 census is the first one to show stabilizing births cohorts at the basis of the agepyramid, with still the possibility of some under-enumeration of the 0-4 age groups, as shown in previous censuses.

The same effect of population growth applies to environment services: access to improved water and sanitation. But it is complicated by urban/rural aspects. It can happen that urban growth consumes most of resources for infrastructures, leaving rural areas deprived, which is actually hindering rural development and increasing the urban drift. – The same probably applies to services in health and education, but water and sanitation indicators are usually disaggregated by urban/rural while it is more rarely so for education and health services which is a real gap in MDGs monitoring and may hide consequences of rapid urbanization.

Education, Child and Maternal Health Indicators

Whenever reduced growth can have favorable effects on education and health services indicators, brain drain depletes the stocks of qualified teachers and nurses. Transition rates from class 1 to 6, or up to form 3 – that some countries have included in their extended MDGs targets-, tend to be low and occur in the frame of many repetitions. A majority of students enter secondary education late by 1 or more years which jeopardize their chances of completing upper secondary education as well as technical education. Low qualification of the population results in difficulty for migrants to migrate through point systems, having to refer to short-term contracts or even migrate undocumented, and face unstable employment and low wages in host countries, which will affect remittances they can send home as well as their poverty status in host country.

IMR and MMR are dependent on coverage and quality of services and therefore they are affected by population growth. However, reducing child and maternal health indicators to very low levels needs high quality of services that is not available due to brain drain of qualified nurses and doctors. Therefore, IMR is stagnating between 10 per 1,000 and 20 per 1,000 in Polynesian migration countries, but it is above 30 per 1,000 in FSM and RMI. As Polynesian countries had already low IMR in the 1990s, achieving the MDGs targets implies rates below 10 per 1,000 that seem difficult to reach.

Gender Indicators

The second impact of migration on population identified above was imbalances in sex ratios. The effect of the sex distribution of migrants, with current increase in feminization of migration, on gender equity and women empowerment has been abundantly commented in the literature (UNFPA, SWOP 2006), but there is not yet a precise and comprehensive quantification of its effects on gender. For women, migration is sometimes a first chance of getting paid employment, but in case of return migration, they may not find similar job opportunities at home. Women left behind, despite remittances from migrant husbands, often need to work which increases the share of women in paid employment and is considered to be empowering them - female labour force participation has also favorable effect on health status of women and children, but it can be detrimental to children education. However, migrants' wives frequently live with relatives or in-law which limits their empowerment. A major benefit of migration for both males and females is behavior change that usually goes with living in a different economic and social environment which can have positive effect on gender equity and increase education of girls. However, this mostly works for the more educated. There is also evidence of migrants' wives living isolated in ghettos and experiencing sometimes even slower social change than in home countries. Social remittances or the dissemination of new ideas through visits of parents/ relatives to host countries and return migration is considered to be mostly missing in the Pacific, despite high mobility of islanders, and the proportion of females in parliamentary seats remains the lowest in the world. Altogether, like urbanization, migration is certainly an opportunity for women. However, this does not occur equally for everyone. There are social groups for which migration does not improve the situation of women and can even have negative effects. This is closely linked with education and economic status and calls to consider the impact of migration on MDG 1: poverty.

Poverty

The central factor in the relation between migration and poverty is remittances. However, labour force participation, a major determinant of remittances, disserves attention in origin as well as in host countries.

The underlying issue of the role of remittances on poverty is whether remittances reduce inequity of income distribution. The development of migration occurs through different phases with different impact on income distribution. At the beginning of migration, there is a 'migration hump': the more educated and wealthy have more opportunity to migrate in the frame of individual migration. In the first phase of migration, the proportion of households with one or more migrants in the lower income quintiles is lower than for higher quintiles and the amount remitted per household is also lower which contributes to increasing the inequality of income distribution. This is the situation shown by Fiji 2003 HIES data for indigenous Fijians whereas migrants and remittances are more evenly distributed by income quintiles for Indians with more ancient and higher migration (Brown 2006). However, this does not mean that there is no impact of remittances on poverty, as some remittances can actually be enough to lift a few households out of poverty, or anyway reduce the depth of poverty. At a later stage, lower social strata access more frequently migration and the impact of remittances on poverty reduction becomes larger. However, *in Tonga, although well advanced in migration process, only 33 per* cent of households in lowest quintile of value of non-land assets had a migrant against 55 per *cent to 75 per cent for higher quintiles* (Brown 2006, Ahlburg 1996). This is not taking into account effects such as increase in commodities, services, food prices and altogether the cost of living, or inflation, that may result from migration and remittances at the national level, pushing higher the poverty lines.

In Asia, the dominant pattern of low qualified labour migration departs somewhat from the phase pattern of migration and remittances on income distribution that prevails in individual migration patterns. Labour migration of unqualified or semi-qualified workers, may speed access to migration of lower (but not the lowest) income people. However, despite high levels of remittances nationally, amounts remitted per household for the poorest segment of the population is certainly low. Whenever, it may contribute to somewhat reducing poverty, it is unlikely to result in much economic development. Countries that rely mostly on unqualified migration show much slower economic growth than those where domestic economy is striving. The former are also more affected by brain drain as local wages are too low to deter qualified workers from leaving and attract return migrants. This is the case of the Philippines, Viet Nam, Indonesia, Bangladesh, Pakistan and Sri Lanka that have much lower GDP per capita than Malaysia and Thailand. Despite low GDP per capita, China and India attract return migrants in the frame of very large differentials in sectoral markets, with high profits being possible in the new technology and export sectors. Pacific island countries have sluggish economic growth and high cost of living which, associated with low wages does not make return migration attractive for nurses and teachers.

Although developed countries are usually not considered in the frame of MDGs (except for Goal 8) as host countries of international migrants, they are concerned by poverty issues as well as population living in slums (MDG 7). Unqualified migrants have often the lowest jobs (3 Ds) and wages in host countries and sending remittances push many migrants' households into poverty. Data from US censuses show that economic position of Pacific migrants, primarily from Samoa, Tonga, Micronesia and Fiji, improved in the 1980s, despite unfavorable macroeconomic conditions, and improved further in the long economic upswing of the 1990s. Part of the improvement in the position of Pacific migrants was due to increases in their human capital: education, work experience and language skills. These gains allowed more workers to acquire white-collar jobs and to increase their earnings (Ahlburg, Song 2006, Ahlburg 2000). Thus, the poverty rate of Pacific Islanders fell from 1.57 times the US rate in 1990 to 1.33 times the US rate in 2000. As regards remittances as well as for the living standard of migrants, migration is obviously more profitable when it consists of qualified workers. Thus, there is a kind of vicious circle with brain drain affecting the production of qualified migrants. Like for other workers, the socio-economic situation of migrants is deeply affected by their education status.

Employment Indicators

MDG indicator 'employment-to-population ratio at ages 15-64' under goal 1 has replaced 'youth unemployment rate' under goal 8. The impact of migration on the old and the new indicator is not much different although it is stronger on the former because migrants are often youth, mostly in the Pacific. As data for the new indicator are not yet available for all countries, we shall consider the old indicator. Migration reduces tensions on the labour market. Whenever, migrants may be employed before migration, their departures make jobs available for non migrants, as long as the economy is not too much affected by out-migration. Moreover, many young migrants never worked before migration, which is typically the case in the Pacific in the frame of preferential migration to New Zealand. However, youth unemployment is still high in these countries, from 12 per cent in Samoa, 30 per cent in Tonga, 35 per cent in FSM and 63 per cent in RMI. This shows that migration without economic growth is not the solution to unemployment. It is still more the case in Asia where emigration countries have actually the highest youth unemployment rates (around or above 20 per cent in Indonesia, Sri Lanka and the Philippines, and migration can only have a marginal effect on unemployment in China and India due to the size of their populations. Economic growth, not emigration, is the best way to reduce pressure on labour markets. This may be a reason why China and India have developed circulation of elites and return migration which contributes to economic growth.

Health and Environment

In origin countries, beside brain drain, migration also reduces the size of labour force in the lowest wage sectors, among which agriculture, reducing agricultural production and implying food imports. In the Pacific, migration and remittances have caused unprecedented extent of changes from the 1960s: modernization of housing, changes in diet and live styles that have resulted in the last decade in epidemic levels of NCDs: cardio-vascular diseases, cancers and diabetes. Dual mortality patterns have appeared in most of Pacific island countries as communicable diseases are still frequent due to poor quality of health services. The high cost of treatment of NCDs uses increasing proportions of the health budget hindering improvement in primary health care. At the end of the day, the impact of migration and remittances on population health as well as on the health budget is probably negative in the mass migration countries of the Pacific. This has led these countries to include NCDs in MDG 6 beside HIV/AIDS, TB and malaria.

Migration and remittances also impact on environment (MDG 7). Changes in consumption patterns result in more toxic waste and pollution affecting air, fresh water resources and coastal reefs. In Polynesia, shortages in agricultural labour force and remittance money have led to inconsiderate use of pesticides that affects coral reefs and reduces sustainability of in-shore fish stocks.

Although poverty is high in RMI and FSM, the increase in automobiles and other household equipments have resulted in large amount of waste that cannot be dealt with in atoll environment and causes chemical pollution that can affect fresh water resources. Lack of adequate sanitation and domestic piggeries in Kiribati have led to high bacteriological pollution of the lagoon in South Tarawa, while pollution from automotives is also increasing. Population density in Ebeye (RMI) reaches above 30,000 per sq.km, it is above 10,000 per sq.km in Majuro urban area and above 8,000 per sq.km in Betio (South Tarawa). These environment issues are also present around cities in other Pacific islands and in Asia.

The relation of HIV/AIDS and TB with migration has already been largely debated, mostly as regards the former. It is true that migration increases risky behavior among lone migrants and results in HIV infection that is transmitted to spouses and partners. Behavior change, women empowerment, safe sex, anti-drug information campaigns are needed to reduce and revert the spread of HIV/AIDS. There is also urgent need to develop relevant information to eliminate inaccurate ideas and discriminative practices against PLHA (people living with HIV/AIDS) as well as migrants altogether. Furthermore, various MDGs indicators can be improved through the many uses of remittances:

- Purchase or in-kinds remittances of telephones, cell phones, computers and internet connection - Pacific countries with the largest expatriate communities are also those with the highest use of internet;
- Payment of education fees for children left behind or other relatives;
- Payment of health expenditures for children, parents or other relatives;
- Use of remittances for collective purposes:

— Infrastructures (wells, improved toilets, sewage, generator/solar power, school books/materials, medicine for health centres), disaster relief,

— Social life: cultural, sports or youth associations, churches..., that can contribute to social change and women empowerment and improve gender equity.

Finally, remittances used for investment or savings contribute to employment generation and economic growth in various ways, with related impact on MDGs indicators. Altogether, whenever migration reduces, if not erases completely, population growth easing increases in services coverage, changes in age structure and dependency due to migration have adverse effects. It delays the advancement in the demographic window of opportunity that is favorable to development. Sectoral shortages in labour force can appear, mostly nurses and teachers for which international demand is high. Brain drain hinders improvement in quality of services causing stagnating infant and maternal mortality and low levels of education and qualification of the population and future migrants, creating a kind of reproduction of marginality.

EMERGING ISSUES

In the frame of the space available in this paper, we shall briefly consider two emerging issues: one relating to bride migration and the other to brain circulation, social remittances, governance and development. Although they are rather already old issues, there has not been much progress done in these areas.

Bride Migration

Bride migration is already an issue in Asia. It is not unknown in the Pacific, mostly in countries where families used to arrange marriages and this now occurs between islands and rim countries. Imbalanced sex ratios at birth have increased in Asia. Female deficit on marriage markets is already felt in China and India, as well as in Korea (Rallu 2006). Given the size of the deficit in China (above 1 million from 2015, or more than a female birth cohort in the Philippines or Viet Nam) and India, it is unlikely that bride migration from neighboring countries can be the solution. However, even limited, bride migration would severely deplete female marriageable cohorts in other countries of the region. This is only one side of the problem. The other side would consist in continued imbalanced sex ratios at birth in these countries based on the belief and hopes that migration will fill the gaps. A continuation of the trends, or stabilization of sex ratios at current levels, would have an impact on population growth and ageing (Attane 2006, Cai, Lavely 2003) as well as consequences on family, society and culture.

Brain Circulation and Social Remittances

The guest worker and highly qualified migration patterns in Asia have moderately developed into brain circulation. Brain drain and sectoral shortages are still increasing. China and India have succeeded in developing brain circulation at a high level of qualification, with PhD students and young researchers studying and working in the US in ICT (often on Chinese or Indian data for social sciences) and returning to teach or work for a few months in universities in their countries of origin. Such pattern is not much developing in other countries for two types of reasons: lack of interest of migrants in returning to work in slowly developing economies, even for short periods, and lack of policy environment that enables such exchange, in the frame of suspicion of administrations, specifically in the Pacific, towards people who have been in contact with other ideas and could try to change the traditional, hierarchical and political systems that prevail in origin countries.

The lack of brain circulation is only part of a larger context that consists in the lack of social remittances in the Pacific. Despite large transnational communities with frequent movements between islands and rim countries, migration has not resulted in much social change in island countries and even only limited change occurs in migrant communities in rim countries. While some migrants leave to escape the constraints of local societies as regards traditional and religious life, most migrants organize their communities around ethnic lines and church leaders in rim countries. Return migrants who set up businesses and do not want to comply with traditional and religious authorities and the custom of gifts to the chiefs, are subjected to various pressures and discriminations and often re-emigrate. Altogether, island countries benefit from remittances, but do not reap the benefits of social and political change that would make business environment more attractive. Migration acts rather well as a safety valve for social problems like unemployment and lack of economic growth so that the need for policy change is not felt, or even is felt as a threat to the traditional system of governance. The low proportion of females in parliamentary seats in the Pacific, actually the lowest in the world, is testimony to the lack of social remittances and related limited women empowerment that does not reach to highest decision making positions.

CLIMATE CHANGE AND MIGRATION

Climate experts assume that global warming may well lead to a global shortage or at least to a displacement of cultivable land. The reasons for this are as diverse as are the impacts of climate change; rising sea levels will lead to more frequent flooding and storms in coastal and delta regions, small island states and low-lying coastal regions could disappear completely as a result. In some regions rainfall will increase significantly, provoking periodic flooding, while in other regions precipitation will quickly decrease, rapidly promoting droughts and desertification. Soil erosion caused by sandstorms and the decline in vegetation will reduce agricultural productivity in these areas — often already low — to a minimum, potentially endangering the food supply for major regions as a whole. The creeping effects of global warming and the associated increase in extreme weather events, as well as the deterioration in living conditions, may give rise to new migration streams.

This policy brief deals with the phenomenon of *environmental migration*. It focuses entirely on the effect of climate change on global migratory movements without neglecting the environmental consequences on the regions of origin and destination. The following paragraphs will firstly contain a comparison of estimates as to how many people will be affected worldwide and an introduction of those areas where climate change is most likely to cause migration. The brief will then examine the two main controversies concerning this phenomenon: the causality relationship between environmental factors and new migratory movements as well as the legal position of the persons concerned. The conclusion emphasises the necessity of extending the protection of people affected by the phenomena of climate change at the international level, even if it cannot be assumed that there is an exclusive causal relationship between climate change and migration.

Estimates

Reliable statistical data cannot be collected as there is not an internationally recognised definition for the phenomenon of climate-induced migration. In addition, estimates are also hindered by the fact that an immediate connection between the consequences of climate change and migration cannot be clearly demonstrated. In the absence of authoritative forecasts, there is a series of estimates based on unsupported assumptions (so-called *guesstimates*). The figures vary depending on which climatic, demographic and social values the estimates are based on. Under favourable conditions, there may be only a slight increase in current migratory movements, but under unfavourable conditions even high estimates appear to be too low.

In 2002 the UNHCR estimated the number of people forced into migration as a result of flooding, famine and other environmental factors at 24 million and later the number of persons displaced internally as a result of natural catastrophes alone at 25 million. The German Advisory Council on Global Change (WBGU) assumes that 10-25 % of all global migratory movements are the result of climate change and its consequences; that would be the equivalent today of an absolute number of 25-60 million migrants. *The United Nations University – Institute for Environment and Human Security*, or UNU-EHS, in

Bonn estimated the number of environmental migrants up to 2010 to be at least 50 million. The Intergovernmental Panel on Climate Change anticipates a total of up to 150 million migrants as a result of climate change by 2050. The United Kingdom's Stern Review bases its estimate on a review of a large number of studies and forecasts and concludes that there are likely to be 200 million environmental migrants by 2050. The figures of Oxford professor Norman Myers are also widespread; he anticipates more than 200 million environmental migrants by 2050.

Affected Areas

In addition to the estimates given above, even the size of the population in areas that will be particularly affected by climate change can provide a useful reference as to the number of people who will be facing special climatic challenges in future and who may possibly regard migration as an alternative. The United Nations standing committee responsible for determining internationally recognised terminology (*Inter-Agency Standing Committee*, IASC) has identified four important scenarios that are likely to trigger migratory movements:

It is a decisive aspect in all scenarios that climate-related migratory movements may take place both within the affected nation states and across international borders, and may be further assigned case-by-case to a continuum of voluntary migration, preventative migration and refugeeism. Such migration may also be either temporary or permanent. Endangered states are deemed in general to be the poorly developed island states (Small Island Developing States, or SIDS), the sub-Saharan states, Asian coastal states, the Polar region, African developing states (Less Developed Countries, or LDC), the least developed countries worldwide (Least Developed Countries, or LLDC), the Near and Middle East, and Central Asia. Depending on the nature of the consequences of climate change, areas affected in line with the IASC scenarios may be divided into the following categories.

AREAS AFFECTED BY SIGNIFICANT, PERMANENT LOSSES IN STATE TERRITORY

This phenomenon comes as a result of rising sea levels and will most probably affect the South Pacific island states in particular (Carteret Islands, Kiribati, the Maldives, the Marshall Islands, Palau, the Solomon Islands, Tokelau, Tuvalu and Vanuatu), which have come to be known as "Sinking Islands", but also low-lying coastal regions in Alaska and the Bay of Bengal. As a result of land losses and the salinisation of coastal regions, some states have already started to permanently relocate inhabitants of their island states, while other countries are not ruling out the possibility of the permanent relocation of all or large parts of their populations. The possibility of relocation to a receiving country or else the founding of new states on uninhabited islands or ceded territories could be considered.

Flood Areas

The rise in sea levels in particular, as well as its hydro-meteorological consequences (increase in periodic floods, tropical storms, coastal erosion, salinisation of coastal waters), represents an important possible inducement for mass-migration. This would affect coastal regions. In addition to small island nations,. According to the Stern Review, by 2080 between 10 and 300 million people will have been affected by the rise in sea level alone, assuming a temperature rise of between 2°C and 4°C.

The IOM estimates that an one-metre rise in sea level would affect 360 000 kilometres of coastline worldwide. Roughly two thirds of the world's population live no further than 100 km from the coast, and areas that lie a maximum of ten metres above sea level alone, the so-called *Low Elevation Coastal Zone* (LECZ), are home to 634 million people – nearly a tenth of the world's current population. Of these, 360 million live in large towns near the coast (in other words, 13% of the global population living in towns). Most of the people in the zone that is affected by rising sea levels live in Asia, Africa and Europe. A current study on the rate of urbanisation in the LECZ recently showed that, alongside the small island states, the densely settled and heavily urbanised deltas and coastal areas in Asia and Africa are particularly exposed to an increased risk of flooding.

Not everyone in the LECZ will have to leave their homes, but rising sea levels could place those in low-lying areas and areas near the coast in acute danger. According to a study carried out by the Potsdam Institute for Climate Impact Research, there are already about 200 million people living in coastal areas that lie less than a metre above sea level.

Thirty of the world's 50 biggest cities lie directly on a seacoast. In the event of a rise of just one metre, according to the study, Egypt's Nile Delta and close to a fifth of Bangladesh (with 35 million inhabitants) would be especially affected, as too would large areas of Suriname, Guyana, French Guiana, the Bahamas, Benin, Mauritania, Tunisia, the United Arab Emirates, Pakistan, India, Vietnam and China. In Europe, an estimated 13 million people would be threatened by a one-metre rise in sea level (especially in the Netherlands and Denmark), including about 3.2 million in the German flood plains. Should sea levels rise by up to one metre, as anticipated, people living in low-lying coastal areas and sea deltas around the world will have hardly any other alternative than to emigrate to other areas.

Drought Zones

Numerous other areas will in future have to contend with a shortage of drinking water due to climate change. The authors of several UN Millennium Ecosystem Assessment studies established that droughts, desertification and the associated decline in agricultural yields are among the strongest factors that will cause people from arid areas to migrate to other regions. The reason for this lies in the far-reaching impact of water shortage, which will bring with it difficulties in supplying drinking water, loss of harvest and health and hygiene problems.

Already today there are more than 1.2 billion people living in regions where there is a shortage of fresh water, i.e. where natural fresh water resources are insufficient to cover the needs of the people living there. This especially affects the northern and sub-Saharan states of Africa, the Near and Middle East, the former constituent republics of the Soviet Union in Central Asia, as well as South East Asia and extensive parts of North China. Some countries in Central and South America also already have to contend with a shortage of water. In all these regions the impact of climate change may lead to longer drought periods, desertification and substantial soil erosion.

REGIONS VULNERABLE TO CONFLICT OVER NATURAL RESOURCES

In addition to emigration movements, the impact of climate change may also lead to conflict over resources. An external WBGU report concludes that, where possible climate-related conflict is concerned, the core regions are in Africa, Asia and Latin America. The climate-induced decrease in cultivable land and water resources affects a population with a growing percentage of youth who already today are likely to migrate into the cities.

This could promote religious, ethnic and civil conflict. The number of inhabitants in regions directly at risk of conflict over resources is approaching one hundred million. If we then add to these the number of inhabitants in areas at indirect risk, the number of potentially concerned persons rises to over one billion.

No matter which trigger for possible environmental migration we examine more closely, those most severely affected will be the small island states as well as the LDCs and LLDCs of Africa and Asia.

But not all of the people living there will migrate for environmental reasons. Infrastructure measures to shore up the coasts, water management plans and new technologies might suffice in a large number of countries and regions to lessen the impact of climate change. Yet, even if only a few percent of the people affected by climate change become environmental migrants, their numbers may reach the scale of the currently estimated refugees and internally displaced people (IDPs) (as at the end of 2008: approx. 42 million).

THE CONNECTION BETWEEN CLIMATE CHANGE AND MIGRATION

The fact that the rise in sea levels or salinisation of coastal areas as *climatic processes*, or hydro-meteorological natural catastrophes as *climatic events*, may trigger migratory movements is not disputed. However, environmental migration does not result froma single cause, but rather incorporates complex interactions of existing social, demographic and political contexts. When considering migratory movements in association with climatic processes or events, therefore, a distinction must be made between climatic and non-climatic migration factors, since migration is not necessarily going to occur for reasons of climatic events alone.

In this regard, adaptation strategies play a decisive role, for a society's vulnerability always results from its particular risk situation in a geographic sense and the efforts such a society makes to adapt. Thus hydro-meteorological catastrophes such as floods or tropical storms only lead to relevant migration phenomena if there have previously been political and social failures to adapt to the specific geographical risk.

In the absence of early warning systems, cross-institutional rescue plans, flood plains or dams, a society's vulnerability in the event of hydro-meteorological catastrophes is increased, as evidenced by the impact of the 2004 seaquake in the Indian Ocean. The tidal waves of the resultant tsunami destroyed entire coastal regions in the Bay of Bengal and South East Asia. At least 165,000 people were swept to their deaths and 1.7 million were left homeless. Some of the main reasons for the devastating impact of the tsunami were the lack of an international early warning and information system as well as the uncoordinated and partially non-existent evacuation of coasts in the affected region. The razing of mangrove forests and elimination of flood zones in coastal areas, as well as their settlement, also contributed to the enormous casualty figures. Not only catastrophes lead to emigration. It is even estimated that the steady degradation of habitable land due to climate change will in future be the most important trigger for international migration. These predictably long-term consequences of climate change already represent a special challenge to the societies that may be affected, for the ecologically induced loss of habitable land is fundamentally "a social problem that can be avoided."

Environmental migration is related to issues that make migration not only necessary, but also attractive, the so-called *pull* factors. These may be of a demographic, social, political or cultural nature. Population pressure, poverty, poor social welfare systems as well as poor governance in states affected by climate change are as decisive triggers for migration as climatic conditions.

At the same time, environmental migration takes place in developing countries in an environment of urbanisation for economic reasons, making it difficult to distinguish environmental migration from "normal" migration in metropolitan catchment areas. Climate change is only one factor in a bundle of factors of varying strength. Migration itself can be interpreted as a means of adapting to the socio-economic and political realities under the conditions of a changing environment. In cases of particularly drastic governmental mismanagement this can mean that a climatic event serves solely as an inducement to migrate, although the main causes are of a political and socio-structural nature.

Environmental migration is therefore not solely based on a simple matter of cause and effect wherein migration is always triggered by climatic conditions alone. It is in fact much more complex than that. If we wish to understand the motives for migratory movement, then previously-existing *pull* factors in particular play a decisive role.

This mutual influence and overlapping of environmental factors with political, social and cultural aspects of migration means that it is not possible to differentiate clearly between voluntary and forced migration, which in turn affects the definition and treatment of people affected by environmental migration.

Categorisation of Affected Persons

There have been numerous attempts to find terminology and definitions for the migration scenarios described above. In addition to the term *environmental migration* used here, there are such expressions as *climate change migration, forced migration* and *environmental refugeeism*. In the English-speaking world the composite term *climigration* is increasingly common. As environmental migration also concerns a mingling of economic and ecological factors and it is virtually impossible to make a clear distinction between these aspects, some authors also refer to *ecomigration*.

The affected people are mostly referred to as *environmental migrants*, but also as *forced climate migrants, environmental refugees* or *environmentally displaced persons*. The terms used for affected people is of decisive importance for categorisation as a migrant or refugee and the resulting consequences with regard to the international obligation to protect or provide for such people. In contrast to migrants, refugees are granted rights by the Geneva Convention concerning aid and services of the United Nations High Commissioner for

Refugees (UNHCR) and may not be deported by receiving states (non-refoulement).

The term *environmental migrant*, coined by the IOM, is finding increasing international acceptance. To facilitate an initial basis for further research and data collection on the phenomenon, the IOM presented a working definition, according to which environmental migrants are "*persons or groups of persons, who, for compelling reasons of sudden or progressive changes in the environment that adversely affect their lives or living conditions, are obliged to leave their habitual homes, or choose to do so, either temporarily or permanently, and who move either within their country or abroad*". This definition seizes on the dimensions considered by the IASC of duration, direction and voluntariness of the migration.

Scientists involved in the European research project EACH-FOR (*Environmental Change and Forced Migration Scenarios*) based their studies on a three-part working definition. They distinguish between *environmentally motivated migrants*, *environmentally forced migrants* and *environmental refugees.* The environmentally motivated migrants differ from the latter two insofar as their change of location is voluntary. The difference between environmentally forced migrants and environmental refugees lies in the fact that forced migrants are subjected to a planned and long-foreseeable, but inevitable migration, whereas climate refugees are forced into sudden emergency migration by catastrophic scenarios. The EACH-FOR working definition does not consider whether in addition to the consequences of climate change there are also social, economic or political inducements to migration, whether the migration is temporary or permanent or whether the migration is only internal or also includes crossing state borders. Like the IOM, the EACH-FOR study picks up on the idea of three levels of duration, direction and voluntariness, but emphasises more strongly than the IOM the possibility of there being mixed causes for migration.

Analogous to the term *Internally Displaced Persons* (IDP), the Norwegian Refugee Council pleads for the descriptive term *Environmentally Displaced Persons* (EDP). This description includes all persons "*who are displaced within their own country of habitual residence or who have crossed an international border and for whom environmental degradation, deterioration or destruction is a major cause of their displacement, although not necessarily the sole one*". The NRC picks up solely on the aspect of direction, i.e. both internally displaced persons and international refugees are included in the definition. The organisation does not consider either the possibility of voluntary migration, such as is allowed for in the IOM definition. The variation of migration triggers are not relevant for the categorisation as a climate migrant, but only the fact that the consequences of climate change are the main trigger of migration.

Controversy has developed in expert circles in particular with regard to the term *environmental refugee*. The reason for this lies in the special legal protection enjoyed by refugees in accordance with the Geneva Refugee Convention (GRC) and additional protocols.

Essentially the question is whether persons affected by climate change should in future be granted refugee protection in accordance with the GRC and its additional protocols. Article 1 A(2) of the convention states that the term refugee shall apply to any person who *"owing to well-founded fear of being persecuted for reasons of race, religion, nationality, membership of a particular social group or political opinion, is outside the country of his nationality and is unable or, owing to such fear, is unwilling to avail himself of the protection of that country; or who, not having a nationality and being outside the country of his former habitual residence as a result of such events, is unable or, owing to such fear, is unwilling to return to it."* As soon as these facts have been proven, the person concerned is granted refugee status.

The UNHCR rejects the use of the terms climate and environmental refugee as a matter of principle, since it fears that the term refugee established by the GRC and its additional protocols could be undermined by the category *environmental refugee*. Other UN organisations that come together under the aegis of the IASC, as well as the IOM, fear that the introduction of the term *environmental refugee* may undermine the established legal instruments for protecting refugees.

The basic conditions for refugee status formulated in the GRC, i.e. the fact of persecution and cross-border migration, would not be met in the case of environmental migration. The impact of climate change does not as yet count as persecution, the majority of the affected persons are internal migrants and therefore still within the protection of their own country. They are therefore less in need of international aid than Convention refugees, according to the UNHCR.

The UNHCR points out that under some circumstances some persons affected by climate-induced migration would meet the conditions for the granting of refugee status in accordance with the GRC. If persecution can be proved for persons fleeing conflict caused by climate problems, then the refugee condition is satisfied. Citizens of the "sinking islands" could also satisfy the GRC conditions if they migrate across borders, because such cases would potentially be a new form of statelessness. If countries of origin were to lose their entire territory, the affected persons could then be treated as stateless and thereby fall under the protection of the Geneva Refugee Convention (GRC) and the attached protocols.

However, the granting of refugee status in the case of the sinking islands scenarios is disputed because it is closely associated with organised or intentional migration. Such intended or tolerated migration can be the result of governmental projects such as the construction of dams or the establishment of flood plains. Both voluntary internal migration (motivated by compensation payments) and forced relocation both within national borders and across international borders occur here.

Essentially, however, the UN Refugee Agency seems to be concerned with preventing the extension of its own mandate due to its already considerable burden at a time when it is financially stretched. It may indeed be one of the organisation's obligations, according to a UNHCR paper, to point out to the international community the gaps in the protection offered to the people concerned, but it is by no means striving to extend its own remit by this means.

In addition, the industrialised nations in particular, which are primarily responsible for climate change, reject the term environmental refugee. Both UN organisations and representatives of industrialised nations constantly refer to the fact that, given the multifaceted and overlapping causes of migration, it is almost impossible to identify the impact of climate change as a main trigger of migratory movements, voluntary or otherwise, with the result that it cannot be proved that any flight is caused primarily by the effects of climate change.

Two scientists working on the EACH-FOR project, Olivia Dun and François Gemenne, counter this argument by pointing out that under the Geneva convention refugees are not anyway required to demonstrate persecution as the main reason for their migration, but rather, the decisive factor for granting refugee status is whether persecution in accordance with Article 1 has actually taken place or not. As soon as any association has been shown between persecution and flight, then according to Dun und Gemenne decision-makers could grant refugee status.

The Norwegian Refugee Council (NRC), which can identify no conclusive definition of the required state of persecution in the UNHCR regulations, also believes that it is entirely possible to recognise climate change as a form of persecution. Thus Paragraph 53 of the *UNHCR Handbook on Procedures and Criteria for Determining Refugee Status* provides for the recognition of refugee status on the basis of "cumulative grounds", not in themselves amounting to persecution, but which, if taken together "produce an effect on the mind of the applicant that can reasonably justify a claim to well-founded fear of persecution". According to the NRC, this concept leaves room for interpretation such that environmental refugees can be protected under the GRC and associated UNHCR regulations.

Moreover, human rights organisations assert that people affected by environmental migration are being robbed of their fundamental right to protection in a situation similar to that of refugees. These people are, by virtue of this, permanent refugees and should therefore also be treated as such. A corresponding category of *environmental refugee* is therefore only logical. Moreover, the migratory movement is a reaction to an externally induced circumstance, similar to a threat or persecution as provided for by the GRC as a condition of refugee status. The organisations therefore plead both for the introduction of the term *environmental refugees* and for an extension to the content of the GRC to recognise such people as "genuine" refugees.

The protection offered to environmental migrants is currently precarious. To date there is still no internationally recognised document requiring that the international community of nations should provide support for environmental migrants in the event that their country of origin is unable to do so. Existing regulations do not oblige international states to take in environmental migrants. Those agreements that do exist can either only be applied in exceptional cases or can be interpreted too broadly to offer reliable protection, or else they are only "can" regulations with no binding effect.

The effects of climate change, alongside other socioeconomic factors, are a trigger for existing and future migratory movements. In practice, however, it will be difficult to make a clear distinction between these triggers in order to identify environmental migration as demanded by some scientists. Specialist literature is divided on the subject of environmental migration. Whereas some scientists deny its existence and speak instead of economic and poverty-driven migration, others regard climate change to be the main reason for migratory movements worldwide.

Environmental migration, like every other social process, takes place within a socio-economic context, so that attempts to draw a precise dividing line between it and other causes of migration, such as war, poverty or climate change, are, in the author's view, doomed to failure from the outset. Nonetheless, it can be assumed that a considerable number of people will be confronted in coming decades with such phenomena as rising seas levels, expanding desert regions and a lack of fresh water. As a result, many of these people will migrate within national borders or across international borders either voluntarily or in flight. Nonetheless, economic, political and cultural aspects of migration must also be considered in order to take account of the complexity of environmental migration. It takes place under the influence of various push- and pull-factors so that answers based on a single cause are not sufficient.

It will be a great challenge in future to decide what status – and consequently what legal status – the affected people are to be granted. International legal norms provide too little protection for environmental migrants, partly due to the absence of any recognition of this new migration phenomenon. The Geneva Refugee Convention (GRC) and its additional protocols only consider some environmental migrants under certain circumstances and therefore do not offer any comprehensive protection. Only a few of today's environmental migrants satisfy the conditions of the GRC, so the majority of persons affected are not currently treated as refugees under current legal conditions. Nor do the legal instruments of nation-states or regions provide environmental migrants with comprehensive protection. It is therefore urgently necessary that regulations should recognise the phenomenon of environmental migration and be adapted to accommodate it. In order not to endanger existing categories, an additional protocol or a new convention appears more meaningful and likely of success than amending the GRC. Furthermore, new regional and national agreements could additionally protect the rights of environmental migrants.

Since the responsibility for climate change rests primarily with the western industrial nations, they are especially responsible for those suffering environmental migration. How far they are ready to meet that responsibility – whether through taking in such people or by providing considerable support in lessening the impact of climate change – will be decisive for the protection of environmental migrants. However, the countries from which environmental migrants originate also have great responsibility towards their citizens and are obliged to do their best to protect their lives. They must take preventive measures to adapt to the consequences of climate change and lessen their impact over both the short and long term.

Climate change presents the international community with great challenges, which can only be overcome if communities work together. Dealing with environmental migration is one of those challenges. If appropriate measures are to be taken, then it is vital to gather additional information about environmental migration. Research into this area should therefore be significantly intensified.

3

Sustainable Development and Environment

Sustainable development has a universal appreciation. At first sight, this is highly positive, as this could signal the entering of a holistic and responsible thinking into the world of politics and society. But as it often happens with other catch phrases that suddenly come into vogue, it fails to be translated into practice, this all the more so because sustainable development can be given several different interpretations. The earliest concept emphasized the need for economic development to be compatible with constraints set by the natural environment, one that satisfies the needs of the present generations without putting in jeopardy the satisfaction of needs of the future. More recently, it has also been stressed that economic development should be compatible with political and social institutions. So a holistic concept of sustainable development has emerged in which economic, ecological, social and political factors need to be simultaneously considered. Participation by individuals, particularly at the community level, is seen as an important means for achieving sustainable development and formulating development goals.

Of course, the concept of 'sustainable development' was not invented in the 1970s or 1980s. Certainly the names of Thomas Robert Malthus and Justus von Liebig have to appear in the early part of the pedigree of this concept. Earlier in this century, social scientists like Thorstein Veblen and economists such as A.C. Pigou drew attention to external costs of economic activities, and in 1950 K. William Kapp published a comprehensive analysis of all the important issues that since the late 1970s have staged a comeback under the name sustainable development. The term 'sustainable development' was probably coined by Barbara Ward (Lady Jackson), the founder of the International Institute for Environment and Development, who pointed out that socio-economic development and environmental protection must be linked. In 1972, a publication about the unsustainability of mainstream development,

The Limits to Growth, triggered enormous fears. The United Nations Conference on the Human Environment, held at Stockholm in 1972, was the first major international discussion of environmental issues. The meeting marked a polarization between the priorities of economic growth and environmental protection. This polarization has dominated the debate between rich and poor countries and between interest groups within countries for many years and - given the results of the Kyoto Climate Conference in December 1997 - is still not fully resolved.

There are legitimate reasons for different perceptions of sustainable development and hence political priorities. Although the most significant ecological issues are of truly global importance, industrial and developing countries still have different problems. For the majority of the people affected by environmental problems in developing countries, lack of sanitation and sewage facilities, polluted drinking water, urban air pollution, shrinking water resources, and eroding topsoil are the most pressing problems. In industrial countries, where such problems have mainly been solved, the public focuses instead on issues such as depletion of the ozone layer as well as the accumulating carbon dioxide in the atmosphere and its potential impact on climatic change.

The World Conservation Strategy promoted sustainable development in 1980, as did The Global 2000 Report to the President prepared under President Jimmy Carter. The concept eventually achieved worldwide recognition and credibility with the publication of 'Our Common Future' (known as the Brundtland Report) in 1987, giving rise to an international consultation process that peaked in the 1992 U.N. Conference on Environment and Development in Rio de Janeiro. Since the early 1990s, understanding of the concept of sustainable development has been widened to include the social dimension and - through the work of Ismail Serageldin and others - has been made more dynamic, so that it involves preserving or enhancing the opportunities of future generations rather than preserving a historically given state of environmental quality or abundance of natural resources. Sustainability is to leave future generations as many opportunities as, if not more than, we have had ourselves.

Over the past few years, substantial progress has been made with regard to the greening of national accounts and hence with regard to measuring the welfare costs of resource depletion and environmental degradation. Comparing the latest set of indicators for sustainable development with the pioneering work of Irma Adelman and Cynthia Taft Morris in the late 1960s illustrates the growing degree of conceptual sophistication. The known policy instruments for environmentally sustainable development have been continuously improved to include subsidy reduction as well as targeted subsidies, environmental taxes, user fees, deposit-refund systems, tradable permits, and international offset systems. A number of elaborate case studies assess the success of different policy instruments.

Politically, the current debate on sustainable development falls into two extremes: One group continues to argue that the end is near, and that only a drastic and widespread change in human behaviour can stop the downward spiral towards self-destruction. The other camp argues that there is no reason to worry, as all trends towards a better life will continue. Experience suggests that the truth lies somewhere in between what could be called environmentalism and techneuphoria.

Most of today's available knowledge suggests that the forthcoming 10 to 30 years are crucial. We know that vital environmental assets, which are not substitutable (like the ozone layer), are being steadily destroyed, and that some of the environmental damages occurring are irreparable (e.g. extinction of species. We still lack a broader understanding of the interdependencies of complex ecosystems, but much of this knowledge will only expand as the natural environment continues to be irreversibly transformed. We would better apply the "precautionary approach", which was brought into the debate by the Rio Declaration on Environment and Development in Principle 15:

> *"Where there are threats of serious or irreversible damage, lack of full scientific certainty shall not be used as a reason for postponing cost-effective measures to prevent environmental degradation."*

SUSTAINABLE DEVELOPMENT IN INDIA

India's final energy demand grows faster than the development of its own national resources. Beyond, one can observe a regular growth in the intensity of polluting energy emissions of the economic activity. This is worsened by the misallocation of resources due to pricing policies, management systems, and more generally, policies that induce a lot of inefficiency and waste. To tackle the long run constraints of the present demand and supply trends, drastic changes in the management of the sector are required. Implementation of reforms began in 1991. Some options exist but a number of bold decisions still have to be taken and implemented to fulfill the energy needs of a population that has now crossed one billion inhabitants. Hence the need to contribute to the debate on sustainable development and scenarios for the twenty-first century In 1972, the then Prime Minister of India, Mrs. Indira Gandhi emphasized, at the UN Conference on Human Environment at Stockholm, that the removal of poverty is an integral part of the goal of an environmental strategy for the world. The concepts of interrelatedness, of a shared planet, of global citizenship, and of 'spaceship earth' cannot be restricted to environmental issues alone. They apply equally to the shared and inter-linked responsibilities of environmental protection and human development. History has led to vast inequalities, leaving almost three-fourths of the world's people living in less-developed countries and one-fifth below the poverty line. The long-term impact of past industrialization, exploitation and environmental damage cannot be wished away. It is only right that development in this new century be even more conscious of its long-term impact. The problems are complex and the choices difficult. Our common future can only be achieved with a better understanding of our common concerns and shared responsibilities.

Sustainable Future

Following are some perspectives and approaches towards achieving a sustainable future:

Finance

Overseas Development Assistance (ODA) is declining. The commitments made by industrialized countries at the Earth Summit in Rio a decade ago remain largely unmet. This is a cause for concern, which has been voiced by several developing countries. Industrialized countries must honour their ODA commitments. The new instruments and mechanisms, e.g., the Clean Development Mechanism, that are trying to replace ODA need to be examined closely for their implications for the developing countries. In view of the declining trend in ODA, developing countries must explore how they can finance their sustainable development efforts, such as by introducing a system of ecological taxation.

Private investment cannot replace development aid, as it will not reach sectors relevant for the poor. Such investments and other mechanisms can at best be additional to, not replacements for, development assistance. Conditions attached to financial assistance need to be rigorously scrutinized, and the assistance accepted only if the conditions are acceptable. Financial support for sustainable development programmes must not be negatively influenced by political considerations external to the objectives of the assistance.

Trade

Trade regimes, specifically WTO, are sometimes in conflict with sustainable development priorities. Imperatives of trade, and the concerns related to environment, equity and social justice however need to be dealt with independently. Environmental and social clauses, which are implicitly, or explicitly part of international agreements must not be used selectively to erect trade barriers against developing countries.

Developing countries will suffer a major trade disadvantage if the efforts to put in place globally acceptable Process and Production Methods (PPMs) are successful. Instead, existing disparities between the trade regimes and multilateral environmental agreements, such as those between Trade Related Intellectual Property Rights (TRIPS) regime and the Convention on Biological Diversity (CBD), should be thoroughly addressed. Mechanisms to resolve such conflicts between multilateral agreements should be set up.

Technology

Developing countries need not follow the conventional path to development with regard to technologies but must use to their advantage the cutting edge technology options now available to 'leapfrog', and put the tools of modern technology to use. Mechanisms must be put in place to make available to developing countries the latest technologies at reasonable cost. Technology transfer must be informed by an understanding of its implications in the social, economic and environmental contexts of the receiving societies. Technologies must be usable by and beneficial to local people. Where possible, existing local technologies must be upgraded and adapted to make them more efficient and useful. Such local adaptations should also lead to the up gradation of local technical skills.

Local innovations and capacity building for developing and managing locally relevant and appropriate technologies must be encouraged and supported. Integrating highly sophisticated modern technology with traditional practices sometimes produces the most culturally suited and acceptable solutions, which also makes them more viable. This trend should be encouraged.

Science and Education

The paramount importance of education in effecting social change is recognized. Mainstream education must now be realigned to promote awareness, attitudes, concerns and skills that will lead to sustainable development. Basic education, which promotes functional literacy, livelihood skills, and understanding of the immediate environment and values of responsible citizenship, is a precondition for sustainable development. Such education must be available to every child as a fundamental right, without discrimination on the basis of economic class, geographical location or cultural identity.

Adequate resources and support for education for sustainable development makers of the potential of education to promote sustainability, reduce poverty, train people for sustainable livelihoods and catalyze necessary public support for sustainable development initiatives. Actions to improve their access to basic and higher education, training and capacity building must support the empowerment of women and girls. The emphasis should be on gender mainstreaming. Greater capacity needs to be built in science and technology through improved collaboration among research institutions, the private sector, NGOs and government. Collaborations and partnerships between and among scientists, government and all stakeholders, on scientific research and development and its widespread application need to be improved.

Population

With India's population crossing a billion in the year 2000, the National Population Policy announced in that year has special significance. Its change in focus from merely setting target population figures to achieving population control through greater attention to socio-economic issues such as child health and survival, illiteracy, empowerment of women, and increased participation by men in planned parenthood, gives it greater breadth and depth, thereby holding forth better promise of achieving its long-term objective of a stable population by mid-century. The official realization, that population is not merely about numbers but about the health and quality of life of people in general and women in particular, must be reinforced and sustained by an informed debate to bring key population issues into ever sharpening perspective at various levels of policy making from the national and state legislatures to local government institutions.

There is need for a better and more widespread understanding that the number of children desired by any couple depends on a large and complexly interrelated number of socio-economic and cultural factors, and that any policy action seeking to control population must seriously take all these variables into account. An important part of empowering women in matters pertaining to population is to explicitly recognize and respect their rights over their bodies and their reproductive behaviour. This recognition must permeate society in general, and religious, judicial and law-enforcement institutions in particular, through continual campaigning and dialogue.

The pursuit of population control must not be allowed to compromise human rights and basic democratic principles. Such compromises are often implicit in the disincentives aimed at controlling family size; in comments on the fertility of particular social groupings; and in the occasional demands to control in-migration to metropolitan areas. It is essential to place these matters in a balanced and rational perspective through informed public discourse supported by the wide dissemination of authentic data.

POVERTY ERADICATION AND SUSTAINABLE LIVELIHOOD

Poverty and a degraded environment are closely inter-related, especially where people depend for their livelihoods, primarily on the natural resource base of their immediate environment. Restoring natural systems and improving natural resource management practices at the grass root level are central to a strategy to eliminate poverty. The survival needs of the poor force them to continue to degrade an already degraded environment. Removal of poverty is therefore a prerequisite for the protection of the environment. Poverty magnifies the problem of hunger and malnutrition. The problem is further compounded by the inequitable access of the poor to the food that is available. It is therefore necessary to strengthen the public distribution system to overcome this inequality. Diversion of common marginal lands to 'economically useful purposes' deprives the poor of a resource base, which has traditionally met many of their sustenance needs. Market forces also lead to the elimination of crops that have traditionally been integral to the diet of the poor, thereby threatening food security and nutritional status.

While conventional economic development leads to the elimination of several traditional occupations, the process of sustainable development, guided by the need to protect and conserve the environment, leads to the creation of new jobs and of opportunities for the reorientation of traditional skills to new occupations. Women, while continuing to perform their tradi-tional domestic roles are increasingly involved in earning livelihoods. In many poor households they are often the principal or the sole breadwinners. A major thrust at the policy level is necessary to ensure equity and justice for them. Literacy and a basic education are essential for enabling the poor to access the benefits offered by development initiatives and market opportunities. Basic education is therefore a precondition for sustainable development. A sizable proportion (about 60% according to some estimates) of the population is not integrated into the market economy. Ensuring the security of their livelihoods is an imperative for sustainable development.

HEALTH AND SUSTAINABLE DEVELOPMENT

Human health in its broadest sense of physical, mental and spiritual well being is to a great extent dependent on the access of the citizen to a healthy environment. For a healthy, productive and fulfilling life every individual should have the physical and economic access to a balanced diet, safe drinking water, clean air, sanitation, environmental hygiene, primary health care and education. Access to safe drinking water and a healthy environment should be a fundamental right of every citizen. Citizen of developing countries continue to be vulnerable to a double burden of diseases. Traditional diseases such as malaria and cholera, caused by unsafe drinking water and lack of environmental hygiene, have not yet been controlled. In addition, people are now falling prey to modern diseases such as cancer and AIDS, and stress-related disorders. Many of the widespread ailments among the poor in developing countries are occupation-related, and are contacted in the course of work done to fulfill the consumption demands of the affluent, both within the country and outside. The strong relationship between health and the state of the environment in developing countries is becoming increasingly evident. This calls for greater emphasis on preventive and social medicines, and on the research in both occupational health and epidemiology. Because of the close link, there needs to be greater integration between them. Basic health and educational facilities in developing countries need to be strengthened.

The role of public health services must give preventive health care equal emphasis as curative health care. People should be empowered through education and awareness to participate in managing preventive health care related to environment sanitation and hygiene. Most developing countries are repositories of a rich resource-based health care. This is under threat, on the one hand from the modern mainstream medicine, and on the other from the degradation of the natural resource base. Traditional medicine in combination with modern medicine must be promoted while ensuring conservation of the resource base and effective protection of traditional knowledge. Developing countries should also strive to strengthen the capacity of their health care systems to deliver basic health services and to reduce environment-related health risks by sharing of health awareness and medical expertise globally.

GDP AND ACCOUNTING FOR SUSTAINABLE DEVELOPMENT

When we calculate GDP, we do not account for the depletion of natural resources. In normal business practice, depletion of capital assets is subtracted from the gross product to get a net product. That's done as "Net Domestic Product" (NDP) for the depreciation of capital assets, but not for environmental assets. The result is that many developing countries have a GDP which is artificially high, because they extract a lot of natural resources, count the sales price in their GDP, but never discount the value of the loss of the natural resources themselves. For example, when Indonesia sells its oil, the selling price of the oil adds into its GDP. Since a large part of Indonesia's economy is based on oil, the government can decide how much their GDP will rise by setting an extraction rate for oil. The politics of that situation is that when an election comes up, the government makes sure the GDP rises, so there's no "recession" to hinder the incumbent. The economics of that situation is that the country should not be used as a model for growth in other developing countries, since those without a large natural resource base cannot do the same tricks with their GDP.

The solution is to account for the depletion of environmental assets, by subtracting from GDP the value that the resource had before it was extracted.

That is, the GDP is decreased by the value lost when the resource was extracted, since that value is no longer owned by the country. The value that a natural resource has when it is unextracted is called its "scarcity rent" (because the unextracted value reflects how scarce the resource is) or its "marginal utility cost" (or MUC). The MUC also applies to environmental amenities other than tangible natural resources, such as clean air, the recreation value of lakes, or the biodiversity provided by forests. When those amenities are lost, by polluting the air, spilling wastes into the lake, or cutting down the forest, the loss in value should similarly be subtracted from GDP. I'll discuss below how to evaluate losses to intangible resources.

If all of the MUCs are summed up, it would represent the total amount of natural resource depletion (NRD, where natural resources includes environmental amenities as well). The NRD is an adjustment to GDP which accounts for the losses of environmental assets, as capital depreciation accounts for the loss of capital assets. The definition of "sustainable development" is that the total amount of the NRD must be reinvested in reproducible capital (assuming that none of the depletion causes irreversible events, such as species extinction).

The MUC for natural resources is the price minus the marginal extraction cost, minus any other distortions which shift the price from its "shadow price," such as monopoly rents, tariff distortions, etc. The NRD for environmental amenities is based on the amount of value lost when the amenity is destroyed. There are a number of means of valuing environmental amenities (discussed below); the NRD can be simply defined as the value before the loss minus the value after the loss. More specifically, the NRD should be the environmental value without the activity causing the loss, minus the value with the activity. The distinction from the before/after definition is that we must also measure what would have happened had we not done the activity which caused the environmental degradation. For example, if we do not extract oil and burn it to create electricity, we would generate power by some other method, which would cause some other environmental degradation. This finer distinction comes into play when doing cost/benefit analyses; it is not so significant here.

To evaluate the losses in environmental amenities, we either look at the effects on marketed goods, or we use a proxy for goods which are not marketed, or we hypothesize a market where none exists. The three examples above correspond to those three levels of market evaluation, as detailed next. Of course, by using the market to evaluate environmental amenities, we are placing a human value on the environment, that is, we are assuming that the only value that matters is that which affects people. "Deep Ecologists" claim an inherent value to environmental amenities independent of their value to humans; that leaves economics and enters ethics, so I won't consider that here.

For tangible natural resources, we measure the value lost because we no longer possess them after we have sold them. The method suggested above is to sum up their scarcity rents, which indicates the value of the unextracted resources. A popular alternative method is to sum up the total asset value at the beginning of the year, subtract it from the total asset value at the end of the year, and consider the difference in value to be the loss due to extraction. The first method is difficult because one must estimate the scarcity rent in markets which are often highly distorted by monopolies (especially for oil), by export subsidies, by import tariffs, etc. Distortionary policies are especially common in developing countries, where natural resources make up a large component of the economy, so this method is most difficult where it is most important to estimate accurately.

The second method is easier to calculate, since one must only tally up the total value of the existing stocks at market value, which is straightforward for tangible resources. The weakness of that method is that if the stock changes during the year, it is interpreted as a gain in value, or a negative NRD. For example, if a new oil reserve is discovered, or if the estimate of an existing reserve size increases, then the total stock of oil may increase even if the quantity of oil extracted was far above the sustainable level. In addition, if the market price of oil rises significantly, the value of the total stock of oil at the end of the year may be higher than the value at the beginning of the year, even if the quantity is much smaller. That problem could be resolved by using economic prices (i.e., with all market distortions removed), but then this method has the same difficulties as the first method. The stock quantity problem is inherent in the method, and makes this method less valid as a measure of sustainability. To measure the value of intangible resources, we cannot use market prices, since markets for the goods themselves do not exist. From the examples above, air pollution uses an actual market price as affected by the unmarketable environmental amenity; lake recreation uses a proxy market price; and biodiversity uses an artificial market price. For each case, I'll describe the method of valuation, which includes both marketable goods and intangible goods.

For clean air, we measure the direct health benefits to the economy, by estimating the number of working days lost when the air is dirty and gets people sick. The marginal improvement in air quality causes a marginal improvement in health, and fewer employees call in sick; the losses from sick days are the cost of air pollution. Since air pollution is typically an urban problem, measuring the human health cost is usually sufficient to evaluate the entire environmental cost. For air pollution or water pollution problems which occur outside of cities, we measure the commercial loss to livestock, fisheries, agricultural production, etc. With acid rain, for example, the cost is primarily in lost revenue to fisheries (because lakes are damaged by acidification), and to farmers (because soil becomes acidified and less productive).

We could also measure the loss in value due to limited visibility, for example, from urban smog. There is no commercial loss due to decreased visibility, and no proxy goods apply, so the "contingent valuation" method (discussed below) would apply. Once again, we only measure the effect on human welfare -- we do not measure the decreased welfare of the livestock or fish, only the value loss that their decreased welfare causes to humans. Deep ecologists have their own welfare decreased by smog, and that welfare loss is captured in the same contingent valuation that measures the value that people place on visibility. A deep ecologist would presumably value clean air more highly than other people, and that value would count -- but the loss in value of clean air to animals themselves, or to the earth itself, or to God, does not count.

For a recreational lake, there is a commercial loss as well as an intangible loss. The loss of cleanliness in a lake creates a direct loss to businesses which use the lake as part of their business, such as boat rentals, fishing guides, campsites, etc. The commercial loss is measured directly, by revenues lost, just as we would measure any other direct loss. The intangible component of the loss is how much value people place on their recreation itself, that is, their utility for their enjoyment of the lake, which they do not pay for in cash terms.

For a lake's intangible recreation value, we use a proxy market, by constructing a demand curve based on what people pay to travel to the lake. The "travel cost method" uses people's willingness to pay to travel to the recreation site as an indication of their valuation. With increasing distance from the recreation site, the cost of getting there increases, for both direct expenses (automobile operating costs, tolls, etc.) and opportunity costs (value of time spent travelling). By graphing the cost of travel versus the number of people from each area, a demand curve is constructed, and then the consumer surplus loss can be estimated for marginal changes in supply of the recreation site. This method has been in use for decades and is the strongest in terms of theoretical foundations, because it uses revealed preferences (actual spending) rather than hypothetical payments.

For a forest's biodiversity value, we use a "contingent valuation" method, where people are surveyed as to how much they'd be willing to pay to preserve the forest. People are shown pictures, for example, of a thriving forest, with many species, versus a managed forest with few species. Then they are asked how much they would be willing to pay, say, in increased taxes, for the difference between the two. A demand curve is thus established again, since many people would be willing to pay a small amount, and decreasing numbers of people would be willing to pay a larger amount. This method is theoretically weak, and expensive to implement, but the US government now accepts contingent valuation in cost/benefit analyses. Problems with contingent valuation begin with the problems of any survey -- the selection process may be biased (only those interested in biodiversity answer the survey); the answers may be strategically high (since the subjects know they don't actually have to pay); the information presented biases the results (different pictures, in the example above, would yield different results). Additional problems arise because it's unclear what is being measured -- people's preference for unspoiled forests, or people's general environmental desires, versus their value for biodiversity per se.

Nevertheless, contingent valuation is the method in use to measure all forms of value which cannot otherwise be measured. Other methods can measure commercial value (losses to business) and use value (losses to recreation). But in many cases, people do not conduct business around environmental amenities, and do not use them personally, but want the amenities to remain accessible so that they have the option to use them, or so that their descendants may enjoy them. People also value unique resources, even in the absence of the option to use them, for their existence itself -- for example, we value the Alaskan wilderness areas because we know that they are pristine, not because we ever intend to go there. Contingent valuation measures the option value, bequest value, and existence value, which no other methods do. Amorphous values such as those sound so vague that they seem to be beyond the scope of economics. But people's willingness to pay for the existence of an Alaskan park, indicates a human welfare increase. The willingness to pay for the existence of environmental amenities is demonstrated by the proliferation of environmental organizations, who earn their living from collecting donations to protect the existence of unusual species, unique natural resources, and other amenities. Amorphous though the values are, good economics must measure the increase in human welfare associated with protecting the environment.

Now let's look at how those techniques apply to GDP calculations. When oil is extracted, or timber removed from a virgin forest, or a lake sacrificed to development, only the value of the commercial goods produced is currently counted in GDP. To measure sustainable development, we should subtract from GDP all of the environmental losses which the development has caused. Extracting oil means that there is less oil in the ground -- the value that the oil would have had, if we had left it unextracted, should be discounted from GDP. Cutting virgin forests means that there is less forest, less recreational opportunity, and less biodiversity -- the loss of those environmental amenities should be subtracted from GDP as well. Development around an undeveloped lake means that the pristineness of the lake is destroyed -- the loss of people's value of the pristineness should be subtracted from the value of the development.

Environmental economists do not say that development in pristine areas should not occur, nor that non-renewable resources should not be extracted. We do say, however, that their full values should be accounted for. Failure to do so means that inefficient activities will occur, that is, activities which ignore part of their costs. Environmental degradation is a large "externality" of development. Internalizing the costs of environmental losses is necessary to properly assess which development is in the interests of people to do. Unaccounted external costs imply that the activity is inefficient, which means that people's net welfare is not maximized. Therefore, ignoring the loss of environmental amenities decreases people's overall welfare.

Accounting for the loss of environmental amenities is a key component of sustainable development. Extracting non-renewable resources at an unsustainable rate is relatively easy to measure and relatively easy to justify limiting the unsustainable activity. Destroying intangible resources is just as unsustainable -- there are a limited number of pristine lakes, and with the development of each one, the value of the remaining ones rises. Scarcity rent applies just as well to intangible amenities as it does to tangible resources. A country which spends its intangible environmental resources should discount the losses due to that spending, or they risk spending the resources unsustainably, and hence the resources will some day be gone.

The definition of sustainable development is that the loss of the limited resources is balanced by the creation of new reproducible capital. Development increases human welfare, but only if the damage caused by development does not cause more losses than gains. The means to determine if development is sustainable is to account for the environmental losses that development causes. Then environment and development will be seen as complements to one another, instead of as adver-saries.

GENERAL PRINCIPLES OF SUSTAINABLE DEVELOPMENT

In addition to the five basic conditions necessary for ensuring sustainable development, five major principles underlie its implementation. These principles, to an even greater extent than the aforementioned conditions, are vital to the definition of sustainable development.

Environmental and Economic Integration

The environment and the economy are obviously very closely related. This link is more than a mere principle; it is a necessity for sustainable development. Various economic tools and policies may promote sustainable development, or at least lead to a more environmentally conscious use of resources. These tools or policies, such as the polluter-payer or consumer-payer approach, may be applied equally to producers, consumers and taxpayers and to enable the market to determine the correct overall cost of using resources. In many instances, however, for the actual value of natural resources to be taken into account, producers and economic agents need to change their attitudes. As a result, tax incentives or other economic tools may be necessary to promote this coming together of the environment and the economy. The integration of the environment and the economy is as advantageous for poorer countries as for rich ones because, if production models adhere to economic and environmental rules, there may be a better balance of comparative production advantages. The result could be a softening of world trade rules whereby poorer countries would be enabled to lay claim to greater economic development.

Certain traditional economic indicators may also assist in assessing the degree to which the economy and the environment are integrated. Particular examples are the gross domestic product and per capita income; global indicators that reflect social aspects (such as the Human Development Index, which includes longevity, education and income); and strictly environmental indicators, such as water quality and land use.

Conservation of Natural Resources

Achieving sustainable development presupposes that we can preserve biological diversity, maintain ecological processes and life support systems and use the world's species and ecosystems in a sustainable manner. Development based on the preservation of natural resources calls for energetic measures that will make it possible to protect the structure, functions and diversity of the natural systems on which life depends. These measures must focus on species and ecosystems as well as on their genetic heritage. Consequently, the limits, on and the capacity for renewal of, natural resources such as soil, wild and domesticated species, forests, pasture and farm land, fresh water and marine ecosystems, must not be compromised. As well, the life of non-renewable resources should be extended by developing and using more effective and cleaner technologies and by encouraging re-use and recycling.

First of all must come changes in the behaviour of individuals and communities and in their attitude to the environment, along with the provision of genuine means for managing it better. New approaches at the state level must then integrate development and conservation of resources on the basis of sufficient information and knowledge and through appropriate legal and institutional instruments. Effort at the international level must be on promotion of the development, and adoption and implementation of conventions and protocols on the environment and natural resources.

Precaution, Prevention and Evaluation

Precaution, prevention and evaluation are the starting points for genuine sustainable development; they must form an integral part of the planning and implementation of every development project. Planners and decision-makers must make it a routine to foresee and provide for the environmental consequences of their projects.

Current environmental protection measures are precautionary; however, in many cases, they are merely a band-aid solution that is not always compatible with the concept of sustainable development, particularly from a long-term perspective. However, the concepts of precaution, prevention and evaluation are difficult to instill because they are often removed from the day-to-day reality and have benefits that will be felt only in the more or less distant future. Forewarned is forearmed, foresight is knowledge and evaluation enables planning: it is imperative that countries and societies adopt these three watchwords so that present development can be transformed into sustainable development.

Cooperation, Partnership and Participation

Achieving sustainable development has become a collective responsibility that must be fulfilled through action at all levels of human activity. Consultation and cooperation in all decision-making are essential to the sustainable management of terrestrial, aquatic and marine ecosystems. It is incumbent upon all states and all nations to cooperate in good faith and in a spirit of partnership in implementing effective strategies to protect, preserve and restore the environment. All must take an active part and do their fair share in accordance with their capabilities and the means at their disposal.

All governments must accept their responsibilities by introducing economic growth policies and programs compatible with the protection of their own environment and that of others. They must ensure the protection of ecosystems of particular importance for agriculture and the way of life of the populations that depend on it. Furthermore, they must facilitate the participation of non-governmental organizations and decentralized or local communities to ensure they can play a greater role in all development-and environment-related activities.

In addition, states must join forces to strengthen international law by adhering to existing environmental conservation and management conventions and protocols and by passing the necessary statutes for their implementation.

They must also promote and develop new agreements and instruments considered necessary to achieving sustainable development.

Cooperation and partnership also presuppose that the richest countries introduce financial and technical assistance measures that will enable the poorer countries to integrate environmental issues more easily into their development programs. The creation of specific environmental protection and restoration funds is certainly worth considering.

The preservation of biological diversity clearly illustrates how interdependent are the "North and South blocs" in the necessary establishment of new partnerships. The main "centres or sources of biological diversity" are situated more particularly in the countries of the South, whereas the major "technological or biotechnological centres" are mainly in the countries of the North. In other words, the countries of the South as well as those of the North must be party to all discussions, solutions and conventions necessary to the achievement of sustainable development. They must all ensure that the measures chosen are suited to the situation of each. The more developed countries will no doubt have to make the necessary efforts to bring about a higher degree of development in the poorer countries and, in particular, the latters' improved access to the most suitable technologies.

Education, Training and Awareness

Safeguarding the environment and achieving sustainable development depend not only on technical and economic matters, but also on changes in ideas, attitudes and behaviour. The direct participation of individuals and communities is essential. All must become fully aware of their environment, know its demands and limits and alter their habits and behaviour accordingly. To this end, countries must develop strategies to better educate, inform and sensitize their populations on environmental matters and sustainable development. For example, ecological and environmental concerns can be integrated into school programs; the awareness of the general public can be raised through extensive information campaigns, particularly through the media; "green" projects can be encouraged in local communities, and training programs can be developed to promote more informed resource management and the use of clean technologies.

SUGGESTIONS FOR SOCIAL DEVELOPMENT

There are several important issues which require detailed deliberation in social development.

The Issues are:-

1. Considering the fast growth of social development sector in 21st century there is need for Government of India to take a proactive role for

multisectoral coordination and convergence of various role players involved in social development for which several Ministries of Govt. of India are required to be sensitized.

2. Need for greater transparency and Accountability in the sector. The world's leading human rights, environmental and social development international organizations such as Action Aid International, AMNESTY International, and Green Peace International, Oxfam International, Save the Children International and World YWCA have today publicly endorsed the first global accountability charter for the non-profit sector to act as responsible players for social development.
   International NGOs play an increasingly influential role. Global public opinion surveys show higher trust in NGOs than in government and business. In addition to an internal desire to be transparent and accountable, the accountability charter also seeks to demonstrate that NGOs deeply value public trust
   In an unprecedented step, international civil society organizations have come together to demonstrate their commitment to transparency and accountability. This initiative builds on the individual, national and sectoral initiatives taken by international NGOs to set standards of accountability and codes of conduct.
3. Need for greater respect, appreciation and support to the social workers. The reason for this they are the key actors for the overall development. Every part of social development sector depends upon their ability and responsibility. Efficient and effective social workers will be made by the concerned NGOs. NPOs, NGOs and other developmental organizations and foundations are needed to respect and appreciate the social workers.
   The lack of social workers and the continuous high turnover of social workers weaken the quality and availability of the service given to clients especially for the poor. The lack of competent personnel may lead to the loss of the basic social rights intended in the constitution. Reasons for the lack of social workers are low pay, lack of leadership and heavy workload.
   In order to improve the availability of social workers and their quality work it is suggested that social workers should have the opportunity to participate in continuing education and supervision of work.
4. Need for further promotive policies of activities self groups, micro credit & market avenues for the productivity self help groups. NGOs can promote the policies for self help groups and micro finance. There is strong relation between the self help groups and micro credit system at present scenario.

In India self help groups are extensively working as primary tools towards poverty alleviation and empowerment. National and state government initiatives, as well as NGOs efforts, have used SHGs to implement poverty alleviation programmes in Andhra Pradesh since 1979. Micro credit is emerged as strong weapon to eradicate poverty through self sustainable with the help of NGOs, micro finance institutions and banks. At present scenario of social development with relation to

5. Need for clear distinction and clarity of Micro Credit & Micro Finance. Mainly SHGs are the prime clients for the Micro and Micro Finance. Banks or financial institutions need to clarify the difference between these two things, because banks are main sources for the micro credit or finance.

   Micro credit and micro fiancé, often used synonymously, is very popular terms in recent developmental activities. This is creating huge confusion and misunderstanding in developmental activities. Professor Muhammad Yunus mentioned this problem with some sarcasm in his address to the International Seminar on Attacking Poverty with Micro Credit, held in Dhaka on 8 and 9 January 2003: The word micro credit did not exist before the seventies. Now it has become a catchword developmental practitioner. In the process, the word now means everything to everybody.

   Micro credit caters commercial needs of poor for enabling them to raise their income levels and improve standard of living. Micro credit means more emphasis on loans while micro fiancé also includes support services where you open up channels for thrift, market assistance, technical assistance, capacity building, insurance, social and cultural programmes. So where there is micro finance is credit plus, there only micro credit is credit.

   Micro credit financial requirements are generally not meant for economic development activities, but for consumptive needs like it education of a child, medicinal requirements etc. Here quantum are quite low, needs are very emergent, and there is hardly any difference between the consumptive purpose and productive purpose.

6. Need for coordinating Agencies & mechanism both at Govt. of India and at various State Govt. levels. Need for coordinating Agencies and mechanism both at Government of India and at various State Government level. Coordinating agencies and mechanism helps in brining the all NGOs together and facilitate for better work in social development with the help of government machineries.

More over, these instruments helps to form networks at all level including international, national, state and local level. This body can also coordinate NGO movements in each country.

7. Need for enabling policies both by centre and states. The social problems of contemporary India are the result of a complex nexus between the factors of exclusion and inclusion rooted in history, values, and cultural ethos. Many of these problems could not addressed by the development strategy launched since independence. Recent policies of globalization have further undermined the role larger societal norms as well as the state apparatus that could counter exclusionary forces. The agenda of social development has remained unfinished, keeping social tensions simmering.

   During the 7th five-year plan, polices were helped to achieved the targeted social development goals, in terms of establishment of social infrastructure, especially in rural areas. The 8th five year plan identified "human development" as its main focus, with health and population control listed as two of six priority objectives. It was emphasized that health facilities must reach the entire population by the end of the 8th plan. The plans also identified people initiative and participation as a key element. With the enactment of the 73rd Constitutional Amendment Act (1992), Panchayati Raj Institutions (PRIs) were revitalized and a process of democratic decentralization ushered in, with similar provisions made for urban local bodies, municipalities and nagar palikas.

   Today, however, in the policy debate, ongoing orthodox economic liberalism is giving way to concerns regarding social consequences of globalization, as it affects the poorest and the marginalized sections of the population. Thus, a number of highly important and far-reaching social policy measures have been brought on to the development agenda, in the form of the right to information act, rural employment guarantee act, the rural health mission among others.

8. Need for capacity building of Govt. Officials, NGOs and gross root level activists and stake holders of social development. Several international and national conferences identify an effective leadership role of the NGOs as social development channels.

   Capacity building can be defined as "development, fostering and support of infrastructure, resources and relationships for NGOs and related systems and services, at Member States, organizational, inter-organizational, and regional and systems levels, contributing to the peaceful, socially distributed and sustainable development of our societies." Capacity-building programmes broaden and strengthen the professional expertise and accelerate progress in the organization activities weather it government or non governmental organization. Capacity building's main goals are to increase the individual capacity of present and future developmental professionals and leaders and to support the development of institutions and programmes all over the world especially in the social development sector.

9. Need for process documentation, Action research, Monitoring & Evaluation and social audit.

   The word "documentation" includes both records and documents. Records are recorded information, regardless of the medium or characteristics, made or received by an organization that is useful in the operation of the organization. Documents explain what an organization plans to do and how it will be accomplished as well as instruct employees how to perform tasks. In this regard professional organizations need to maintain quality and effective documentation along with the action research, monitoring and evaluation and social audit.

   Action research is very important component for the effective functioning of the organization. Action research finds out the achievement and failures of the organizations activities. With the help of action research findings, NGOs and other developmental organizations can reform or modify their on going developmental projects, if there are any insufficient methods or policies, social audit is also one of the important elements for the social development. NGOs are the prime players to enhance the awareness about the social audit within the NGOs and other governmental organizations.

10. Need for further promotion of participatory approaches of the development.

Participatory approach of the development facilitates the local communities play a central role in the planning, implementation and funding of activities within participatory developmental programmes. The exact composition of any given programme should be determined in conjunction with them. It is important to ensure that programme activities:

1. do not provoke conflict between resource users (where conflict is unavoidable, conflict resolution mechanisms should be specified early on);
2. do not further isolate marginal households (that may not be able to participate in activities which demand a labour or financial contribution);

3. do not undermine viable indigenous soil and water conservation techniques;
4. are informed by an understanding of existing management practices (e.g. they do not immediately promote group activity if there is no history of communal working);
5. are feasible given current capacity within the community and external organizations; and
6. take into account underlying climatic, hydrological, soil and land use characteristics.

   Participatory approaches are more important to succeed the developmental programmes. Participatory approach also enables the social capital for social development.

11. Need for sharing of knowledge, innovative approaches, and people centred approaches. Here, sharing of knowledge is very important component in every aspect of development field. This concept of sharing of knowledge leads to innovative methods in development sector with development organizations. People centred approach, at present, is very much needed in the field of social development. These above components considered as critical elements for speedy social development. NGOs and other developmental organizations need to develop an innovative approach for social development with the help of sharing of knowledge concept. To get access of sharing of knowledge and to develop innovative approach, NGOs and other developmental organizations need to get extensive trainings on various issues such as Organizational Development, Capacity Building of Organizations and other related to social development. Moreover, NGOs are considered by international developmental organizations and banking as the prime actors for social development.
12. Need for savings, economy of resources and no cost, low cost and cost effective approaches. There is interrelation between the above concepts such as savings, economy of resources and no cost, low cost and cost effective approaches in social development. Developing the society with all minimum needs required the tremendous savings in all sections of the society. Especially in rural areas, women self help groups savings are reached at maximum level. This saving came from only marginalized sections of the society. In this scenario, NGOs and other developmental organization's efforts utilized cent percent with no cost and low cost effective approaches. Thus, for overall social development need the above concepts. Here, NGOs and other voluntary or developmental organizations need to enhance the awareness on above issues for better social development.

13. Need for exchange visits, study tours, & National and International exposure. Exchange visits and study tours at national and international exposure will broaden and strengthen the organizational capacity in various aspects such as management skills, developmental methods and interaction with developmental professionals in the field of national and international and so on. Moreover, tours and visits also give knowledge in the different areas of social development from different nations especially from developed countries. The developed nation's technological knowledge will help in faster the social development in developing countries.
14. Need for planning by NGOs working in the sector. As earlier stated by international developmental organizations, NGOs are the critical players in social development. Because, NGOs works in the every aspect of society and they works from grass roots level to international level. So, NGOs can draw the plans for social development effectively.
15. Need for introduction of managerial inputs to the NGOs working in the sector. For the effective programme implementation and policy-making, NGOs need to acquire the knowledge of effective managerial inputs for the NGOs in the different areas of social development. In the present scenario of social development, every programme of poverty alleviation in rural as well as urban areas and infrastructure development such as roads, drinking water, schools and health centres etc., are implemented and monitored by NGOs. Thus, for effective implementation or benefits reaching to the poor are depends up on NGOs effectiveness.
16. Need for sustainable development approaches. Need for collaboration among peoples institutions, NGOs and Government Institutions. The international community has recognized the vital importance of cooperation between government agencies and non-governmental organizations (NGOs) in addressing the social and human development related issues. The importance of government-NGO cooperation was stressed in the recent global and regional conferences concerning the social development. To respond to these mandates, international developmental organizations formulated the process of government-NGO collaboration for social development especially in the area of poverty alleviation programmes. Moreover, World Bank also is stressed on the Government and NGO collaboration for rapid social development in recent international conferences.
17. Need for Multisectoral coordination and convergence of various role players in the sector. Multi sector coordination and convergence is the

effective method for rapid social development. Governments across the world are grappling with appropriate policies to optimize the benefits associated with convergence through multi sectoral coordination. Convergence has emerged as a global phenomenon as a result of digitization which has allowed traditionally distinct services to be offered across interchangeable platforms. These technological trends have been accelerated by the liberalization of markets allowing for the social development.

18. Need for integrated Micro Planning at village level and Holistic development. Integrated Micro planning is the most important concept for the developmental and governmental organizations. In this aspect, every resource of the village utilized for the overall development of the village through the micro planning. In this process, natural resources and human resources utilized for overall development with the help of NGOs and government organizations.
19. Need for creation of the Model Villages. Model villages especially created for the utilization of information and communication technology. In the previous years, ICT was utilized for the development of corporate industries and other large size multi national corporations. Now, in the globalized era, ICT is reaching to the every corner of the globe including villages. If ICT need to utilize properly in village development, villages need to be developed as model villages so as to accept the access of developed technologies. NGOs and other developmental organizations need to be developed the facilities in rural areas with coordination with the government organizations.
20. Need for non exploitation, lack of corruption, and lack of hypocritic approaches by a few black sheep in social development. As late Prime Minister of India, Rajeev Gandhi noticed that exploitation and corruption are the major hurdles for the path of India's development. This is indicating that the high intention and impact in the social development. NGOs and other developmental organizations and charities and every social group are need to work for the non exploitation, lack of corruption, and lack of hypocritic approaches for the better society and social development.

The inter-relationship between the various activities has to be emphasized and the necessary coordination assured both in the Central Government; and in the States. One aspect of this coordination would be to secure that legislation relating to social problems follows broadly similar principles.

In cases where grants-in-aid are given by a State authority to a private agency, it is desirable to lay down general directions for improving the content of the programmes and their administration. A measure of supervision and inspection should also be provided in order to maintain standards of efficiency. A major responsibility for organizing activities in different fields of social welfare, like the welfare of women and children, social education, community organization, etc., falls naturally on private voluntary agencies. These private agencies have for long been working in their own humble way and without adequate State aid for the achievement of their objectives with their own leadership, organization and resources.

Any plan for the social and economic regeneration of the country should take into account the service rendered by these private agencies and the State should give them the maximum cooperation in strengthening their efforts.

Public cooperation, through these voluntary social service organizations, is capable of yielding valuable results in canalizing private effort for the promotion of social welfare.

One of the most important tasks of the State is to conduct a survey of the nature, quality and extent of service rendered by voluntary agencies in different parts of the country, to assess the extent of financial and other aid that they are in need of in order to develop their programmes of work, and to coordinate their activities.

CHANGING COUNTRIES ECONOMIC POLICIES

Environmental damage in industrialized and developing countries can be reduced with an existing package of measures that could be implemented without reducing the standard of living of the people that would be affected by them. Natural resources, such as air and water are being regarded as "free goods." Their costs are "externalized," i.e. they are paid by society in the form of damage to the ecosystems. Prices and market mechanisms must be adjusted to reflect environmental costs. They have to become an integral part of business calculations. Society must establish adequate prices for the use of goods held in common, water, atmosphere, air, and land. Governments will have to move beyond the traditional command-and-control regulatory approach and use more market-oriented solutions, which offer incentives and rewards to those who continuously innovate and improve in the area of environmental impacts. This work must be based both on the best available scientific evidence and on people's preferences and choices. When resources are priced properly, resource-intensive goods will become more expensive - and hence less attractive to consumers. Competition encourages producers to make the use of such goods cost-effective, i.e. to minimize it. To the extent that waste represents resources that have escaped from a production system, concern for costs will also encourage producers to minimize waste, especially when they pay to control it or are made liable for the damage it causes. Companies, which take their responsibility towards the environment seriously and develop better products and processes, will have a competitive edge over others. The competition inherent in open markets is the primary driving force for the creation of ecologically sound technology. Hence with an internalization of environmental, with adequate legal regulations, improved resource management (particularly with greater energy efficiency), but also with a change in behaviour patterns and different ways of defining "wealth" and "living standard", a change in our ecological (and social) course is feasible over the next 15-25 years.

The Dutch Advisory Council for Research on Nature and Environment came up with concrete data about the desired quantitative reduction in the use of eco-capacity. If a sustainable level of the use of eco-capacity should be achieved, the use of fossil fuels, metals and renewable resources as well as CO2 emissions, acid deposition and deposition of nutrients and metals must be reduced by 70 to 99% over the next 45 years.

As the use of eco-capacity was and is unequally distributed among the world population, the North will have to relinquish part of its claim on the ecocapacity for the benefit of the South - this not only for reasons of equity but also from the point of view of international security. The study "Sustainable Netherlands" proposes a programme of action for the sustainable development of the Netherlands that works on similar value premises and comes up with comparable conclusions. Both studies underline the point, that time is an important factor in the sense that an early onset of action will lead to less social and economic friction.

But then, how will this be done politically? In democratic societies, politicians get elected and stay in power by voters living today. So far, it has been impossible to win elections by burdening political constituencies today with costs, shrinking options if not bans for the sake of future generations. Hence, as the Club of Rome puts it,

> *"Governments give priorities to politically useful short-term solutions and systematically neglect the longer-term perspective. As a consequence of such legacies of neglect, problems tend to become compounded and governments fall into a rhythm of crisis government".*

There is a tendency to treat sustainable development as merely a variation of the prevailing approaches to development and to see sustainability as a goal that can be attained through making adjustments to the standard development models. Enlightened heroism on the side of professional politicians to do the right thing-and to run a high risk of being ejected from power - is not a widespread quality. Thus, also here, "development from below" is necessary.

ENVIRONMENTAL ECONOMICS

Environmental economics is a subfield of economics concerned with environmental issues. Quoting from the National Bureau of Economic Research Environmental Economics program: "Environmental Economics undertakes theoretical or empirical studies of the economic effects of national or local environmental policies around the world. Particular issues include the costs and benefits of alternative environmental policies to deal with air pollution, water quality, toxic substances, solid waste, and global warming."

Topics and Concepts

Central to environmental economics is the concept of market failure. Market failure means that markets fail to allocate resources efficiently. As stated by Hanley, Shogren, and White (2007) in their textbook *Environmental Economics*: "A market failure occurs when the market does not allocate scarce resources to generate the greatest social welfare. A wedge exists between what a private person does given market prices and what society might want him or her to do to protect the environment. Such a wedge implies wastefulness or economic inefficiency; resources can be reallocated to make at least one person better off without making anyone else worse off." Common forms of market failure include externalities, non excludability and non rivalry. Externality: the basic idea is that an externality exists when a person makes a choice that affects other people that are not accounted for in the market price. For instance, a firm emitting pollution will typically not take into account the costs that its pollution imposes on others. As a result, pollution in excess of the 'socially efficient' level may occur. A classic definition is provided by Kenneth Arrow (1969), who defines an externality as "a situation in which a private economy lacks sufficient incentives to create a potential market in some good, and the nonexistence of this market results in the loss of efficiency." In economic terminology, externalities are examples of market failures, in which the unfettered market does not lead to an efficient outcome.

Common property and non-exclusion: When it is too costly to exclude people from accessing a rivalrous environmental resource, market allocation is likely to be inefficient. The challenges related with common property and non-exclusion have long been recognized. Hardin's (1968) concept of the tragedy of the commons popularized the challenges involved in non-exclusion and common property. "commons" refers to the environmental asset itself, "common property resource" or "common pool resource" refers to a property right regime that allows for some collective body to devise schemes to exclude others, thereby allowing the capture of future benefit streams; and "open-access" implies no ownership in the sense that property everyone owns nobody owns. The basic problem is that if people ignore the scarcity value of the commons, they can end up expending too much effort, over harvesting a resource (e.g., a fishery). Hardin theorizes that in the absence of restrictions, users of an open-access resource will use it more than if they had to pay for it and had exclusive rights, leading to environmental degradation.

Public goods and non-rivalry: Public goods are another type of market failure, in which the market price does not capture the social benefits of its provision. For example, protection from the risks of climate change is a public good since its provision is both non-rival and non-excludable. Non-rival means climate protection provided to one country does not reduce the level of protection to another country; non-excludable means it is too costly to exclude any one from receiving climate protection. A country's incentive to invest in carbon abatement is reduced because it can "free ride" off the efforts of other countries. Over a century ago, Swedish economist Knut Wicksell (1896) first discussed how public goods can be under-provided by the market because people might conceal their preferences for the good, but still enjoy the benefits without paying for them.

Valuation

Assessing the economic value of the environment is a major topic within the field. Use and indirect use are tangible benefits accruing from natural resources or ecosystem services. Non-use values include existence, option, and bequest values. For example, some people may value the existence of a diverse set of species, regardless of the effect of the loss of a species on ecosystem services. The existence of these species may have an option value, as there may be possibility of using it for some human purpose (certain plants may be researched for drugs). Individuals may value the ability to leave a pristine environment to their children. Use and indirect use values can often be inferred from revealed behaviour, such as the cost of taking recreational trips or using hedonic methods in which values are estimated based on observed prices. Non-use values are usually estimated using stated preference methods such as contingent valuation or choice modelling. Contingent valuation typically takes the form of surveys in which people are asked how much they would pay to observe and recreate in the environment (willingness to pay) or their willingness to accept (WTA) compensation for the destruction of the environmental good. Hedonic pricing examines the effect the environment has on economic decisions through housing prices, travelling expenses, and payments to visit parks.

Solutions

Solutions advocated to correct such externalities include:

- *Environmental regulations*. Under this plan the economic impact has to be estimated by the regulator. Usually this is done using cost-benefit analysis. There is a growing realization that regulations (also known as "command and control" instruments) are not so distinct from economic instruments as is commonly asserted by proponents of environmental economics. E.g.1 regulations are enforced by fines, which operate as a form of tax if pollution rises above the threshold prescribed. E.g.2 pollution must be monitored and laws enforced, whether under a pollution tax regime or a regulatory regime. The main difference an environmental economist would argue exists between the two methods, however, is the total cost of the regulation. "Command and control" regulation often applies uniform emissions limits on polluters, even though each firm has different costs for emissions reductions. Some firms, in this system, can abate inexpensively, while others can only abate at high cost. Because of this, the total abatement has some expensive and some inexpensive efforts to abate. Environmental economic regulations find the cheapest emission abatement efforts first, then the more expensive methods second. E.g. as said earlier, trading, in the quota system, means a firm only abates if doing so would cost less than paying someone else to make the same reduction. This leads to a lower cost for the total abatement effort as a whole.
- *Quotas on pollution*. Often it is advocated that pollution reductions should be achieved by way of tradeable emissions permits, which if freely traded may ensure that reductions in pollution are achieved at least cost. In theory, if such tradeable quotas are allowed, then a firm would reduce its own pollution load only if doing so would cost less than paying someone else to make the same reduction. In practice, tradeable permits approaches have had some success, such as the U.S.'s sulphur dioxide trading program or the EU Emissions Trading Scheme, though interest in its application is spreading to other environmental problems.
- *Taxes and tariffs on pollution/Removal of "dirty subsidies"*. Increasing the costs of polluting will discourage polluting, and will provide a "dynamic incentive", that is, the disincentive continues to operate even as pollution levels fall. A pollution tax that reduces pollution to the socially "optimal" level would be set at such a level that pollution occurs only if the benefits to society (for example, in form of greater production) exceeds the costs. Some advocate a major shift from taxation from income and sales taxes to tax on pollution-the so-called "green tax shift".
- *Better defined property rights*. The Coase Theorem states that assigning property rights will lead to an optimal solution, regardless of who

receives them, if transaction costs are trivial and the number of parties negotiating is limited. For example, if people living near a factory had a right to clean air and water, or the factory had the right to pollute, then either the factory could pay those affected by the pollution or the people could pay the factory not to pollute. Or, citizens could take action themselves as they would if other property rights were violated. The US River Keepers Law of the 1880s was an early example, giving citizens downstream the right to end pollution upstream themselves if government itself did not act (an early example of bioregional democracy). Many markets for "pollution rights" have been created in the late twentieth century—see emissions trading. The assertion that defining property rights is a solution is controversial within the field of environmental economics and environmental law and policy more broadly; in Anglo-American and many other legal systems, one has the right to carry out any action unless the law expressly proscribes it. Thus property rights are already assigned (the factory that is polluting has a right to pollute).

Relationship to other Fields

Environmental economics is related to ecological economics but there are differences. Most environmental economists have been trained as economists. They apply the tools of economics to address environmental problems, many of which are related to so-called market failures—circumstances wherein the "invisible hand" of economics is unreliable. Most ecological economists have been trained as ecologists, but have expanded the scope of their work to consider the impacts of humans and their economic activity on ecological systems and services, and vice-versa. This field takes as its premise that economics is a strict subfield of ecology. Ecological economics is sometimes described as taking a more pluralistic approach to environmental problems and focuses more explicitly on long-term environmental sustainability and issues of scale.

Environmental economics is viewed as more pragmatic in a price system; ecological economics as more idealistic in its attempts not use money as a primary arbiter of decisions. These two groups of specialists sometimes have conflicting views which may be traced to the different philosophical underpinnings.

Another context in which externalities apply is when globalization permits one player in a market who is unconcerned with biodiversity to undercut prices of another who is-creating a "race to the bottom" in regulations and conservation. This in turn may cause loss of natural capital with consequent erosion, water purity problems, diseases, desertification, and other outcomes which are not efficient in an economic sense. This concern is related to the subfield of sustainable development and its political relation, the anti-globalization movement. Environmental economics was once distinct from resource economics. Natural resource economics as a subfield began when the main concern of researchers was the optimal commercial exploitation of natural resource stocks.

But resource managers and policy-makers eventually began to pay attention to the broader importance of natural resources (e.g. values of fish and trees beyond just their commercial exploitation;, externalities associated with mining). It is now difficult to distinguish "environmental" and "natural resource" economics as separate fields as the two became associated with sustainability. Many of the more radical green economists split off to work on an alternate political economy.

Environmental economics was a major influence for the theories of natural capitalism and environmental finance, which could be said to be two sub-branches of environmental economics concerned with resource conservation in production, and the value of biodiversity to humans, respectively. The theory of natural capitalism (Hawken, Lovins, Lovins) goes further than traditional environmental economics by envisioning a world where natural services are considered on par with physical capital.

The more radical Green economists reject neoclassical economics in favour of a new political economy beyond capitalism or communism that gives a greater emphasis to the interaction of the human economy and the natural environment, acknowledging that "economy is three-fifths of ecology"-Mike Nickerson.

These more radical approaches would imply changes to money supply and likely also a bioregional democracy so that political, economic, and ecological "environmental limits" were all aligned, and not subject to the arbitrage normally possible under capitalism.

Professional Bodies

The main academic and professional organizations for the discipline of Environmental Economics are the Association of Environmental and Resource Economists (AERE) and the European Association for Environmental and Resource Economics (EAERE). The main academic and professional organization for the discipline of Ecological Economics is the International Society for Ecological Economics (ISEE).

CONSENSUS FORMATION OF BUSINESS AND INDUSTRY

Throughout this process of consensus formation, business and industry exerted a structuring influence. They succeeded in making their view hegemonic, and ended up being considered post-Rio as a major social actor providing solutions to the global ecological crisis. As influential economic agents, transnational corporations (TNCs) have activities that directly impact on the situation of the environment. TNCs have been a constant target of NGOs, which point out their preponderant role in environmental degradation. Several public campaigns and boycotts have been organized to draw the public's attention on the issue and force TNCs to comply with legislation, adopt higher environmental standards or change production processes.

On the issue of tropical deforestation for example, NGOs have pointed out that corporations such as British Petroleum, Shell or Mitsubishi bear a large responsibility for forest devastation worldwide. Already in 1989, The Sunday Times directly accused British Petroleum and Shell of contributing to the depletion of the Amazonian rainforest in Brazil.

More recently, the Rainforest Action Network (RAN) accused Mitsubishi, together with its subsidiary Meiwa, of being "the greatest corporate threat to the world's tropical, temperate and boreal forests." RAN accuses Mitsubishi of illegal logging, transfer pricing, tax evasion, violations of pollution standards, anti-trust activity, violation of native land claims, and employment of illegal aliens. Yet despite evidence of the role of corporations in environmental degradation, the issue was scarcely discussed and questioned during the UNCED process. There is, it is true, a chapter in Agenda 21 dedicated to the role of business and industry. Yet the document does not in any way blame business for its major contribution to the ecological crisis. Agenda 21 contents itself with providing guidelines to firms in order to help them improve their environmental records.

But this is not to say that business and industry were absent or uninterested in the negotiation. On the contrary, large corporations were very active in the UNCED process, and even before it. As early as 1984 a World Industry Conference on Environmental Management (WICEM I) had been organized in France to recommend actions to include environmental concerns in industry planning. WICEM II, which took place in 1991, adopted sustainable development as its main axiom. The corporations agreed that there should be convergence, and not conflict, between economic development and environmental protection, and launched the Business Charter for Sustainable Development. In 1990, the Business Council for Sustainable Development (BCSD) was created under the chair of the Swiss industrialist Stephan Schmidheiny, personal friend of Maurice Strong (UNCED's Secretary General) and his special adviser for business and industry during the UNCED process. The BCSD was created as a group of 48 chief executive officers of corporations from all regions of the world, some of them with a rather negative environmental record, including Chevron, Volkswagen, Nissan, Nippon,

Mitsubishi, Dow, Shell, CVRD, Aracruz, and Axel Johnson. The BCSD was closely involved in the preparation of the Conference, and, through Strong, had special access to UNCED's Secretariat. As a result, after Rio, corporations became "partners in dialogue," and their vision of sustainability became the dominant vision. According to Chatterjee and Finger (1994), corporations shaped the very way environment and development are being looked at: business and industry's worldview came out of Rio as the solution to the global environmental crisis and no longer as its cause.

In the words of the BCSD, "the cornerstone of sustainable development is a system of open, competitive markets in which prices are made to reflect costs of environmental as well as other resources. When viewed within the context of sustainable development, environmental concerns become not just a cost of doing business, but a potent source of competitive advantage. Enterprises that embrace the concept can effectively realize the advantages in more efficient processes, improvements in productivity, lower compliance costs, and new market opportunities." Thus, by creating competitive advantages, environmental concerns can provide corporations with new market opportunities and be the source of new profit. Finally, business sees the new era of global development as the era of market efficiency. "It is time for business to take the lead," says Schmidheiny; "change by business is less painful, more efficient, and cheaper for consumers, for governments, and for business themselves. By living up to its responsibilities, business will be able to shape a reasonable and appropriate path toward sustainable development". The ecological crisis perceived in fact by business not as a real crisis but rather as a set of adverse and controllable side-effects of development. Hence it is to be solved via increased efficiency which is to be achieved not through government regulation, but through open markets with a new concern for internalizing externalities.

Today, the BCSD has become the WBCSD (World Business Council for Sustainable Development), under the chair of Börn Stigsen. It now has 125 members representing companies such as British Petroleum, Ciba Geigy, Nestle, Monsanto and the Western Mining Corporation. The WCSD is said to have led industry input into the UN Commission for Sustainable Development and UNCED's 1997 review, revealing the emergence of corporate environmentalism as a driving force of global environmental management.

As stressed by Karliner (1997), after Rio, global corporate environmentalism has helped build a public image of transnational corporations as the world's responsible global citizens, setting the terms of the debate along lines favourable to their interests. In the process, corporate environmentalism has partially neutralized efforts - ranging from popular environmental movements to intergovernmental treaties and conventions - that pose a threat to their activities. While before Rio the environmental movement used the system to advance its goals, now the system has appropriated the environmental discourse and is using the environmental movement. This new strategy has meant increased efforts by corporations to increase cooperation with other environmental actors, in particular with the environmental movement. As noted by Bryant and Bailey (1997, 120), TNCs have sought to cultivate links with moderate NGOs in order to neutralize the threat posed to business from environmentalists. Actually, some NGOs today depend on TNCs for financial support. Stauber and Rampton (1995) observe that this process of funding NGOs and cooperating with them is part of a larger attempt to divide-and-conquer the NGO sector by winning support among moderate NGOs while attacking radical NGOs which campaign against TNCs' activities. Moderate NGOs and TNCs became partners in the international environmental establishment and now work together in the system of global environmental governance.

From Rio 92 To New York 97: The Rise And Fall Of "Global Environmental Management" UNCED's Review Five Years after Rio Five years after Rio, as foreseen at UNCED, the review of UNCED's implementation culminated with the June 1997 New York Summit, often referred to as "Earth Summit II." Earth Summit II's official name is UNGASS, United Nations General Assembly Special Session. During UNGASS, five years of work of the Commission on Sustainable Development (CSD) were presented, including a report by the Secretary-General assessing the progress achieved in the implementation of Agenda 21 and recommendations for future action and priorities.

UNGASS was carried out at the highest level of political representation - Heads of State and Governments - and, as UNGASS itself said, aimed to "re-energize our commitment to further action on goals and objectives set out by the Rio Earth Summit." A new energy was indeed necessary: the main outcome of the meeting was the public recognition of the failure of international efforts to promote long-term sustainability. Yet it only adopted a document, the "Program for the Further Implementation of Agenda 21," and did not produce a political statement or binding commitments needed to reverse unsustainable trends. The text acknowledges that, five years after UNCED, the state of the global environment has continued to deteriorate, and reviews the situation in all areas of action.

It notes progress in institutional development, international consensus-building, public participation and private sector actions, which have allowed some countries to curb pollution and slow the rate of resource degradation. Yet, overall, trends are worsening, polluting emissions have increased, and marginal progress has been made in addressing unsustainable production and consumption patterns. Inadequate and unsafe water supplies are still aggravating health problems, the situation of fragile ecosystems is still deteriorating, and non-renewable resources are used at an unsustainable rate. Despite progress in material and energy efficiency, the report concludes that overall trends remain unsustainable. The document then reviews progress in all sectors and issues, inter alia, fresh water, oceans and seas, forests, energy, transport and atmosphere. Finally, it recommends means of implementation and adopts a program of work of the CSD for the next five years, with a commitment to ensure that the next comprehensive review of Agenda 21 in 2002 demonstrates greater measurable progress in achieving sustainable development.

Interestingly enough, all these trends are examined within the framework of economic globalization. The very assessment of progress made since UNCED starts by highlighting that the five years elapsed since then have been characterized by the accelerated globalization of interactions among countries in the areas of world trade, foreign direct investment and capital markets. The document recognizes the unevenness of the globalization process, stressing that marginalization and income inequality is increasing in some countries as well as within countries and that unemployment has worsened in many countries.

Yet it is believed that globalization presents new opportunities and challenges. The report notes that a limited number of developing countries have been able to take advantage of those trends, attracting large inflows of external private capital and experiencing significant export-led growth and acceleration of growth in per capita gross domestic product.

The view is thus that all countries could take advantage of the globalization trend. It is not perceived that only a few countries, due to specific conjunctural conditions, including interest rates and the monetary situation for example, can attract the volume of FDI necessary to feed the high growth rates praised in the document. The conceptual link with economic globalization appears as somehow flawed. It is not mentioned that significant export-led growth and the acceleration of growth in per capita GDP, if not controlled by an effective system of environmental protection, might be responsible for the worsening of overall trends for sustainable development.

In addition, though the text perceives unsustainable patterns of production and consumption as the major cause of continued deterioration of the global environment and observes that unsustainable patterns in the industrialized countries continue to aggravate the threats to the environment, only very vague actions and guidelines are adopted to change them, such as recommending the internalization of environmental costs, developing indicators, promoting efficiency, information, technology, and the role of business in shaping more sustainable patterns of consumption.

No binding commitment to deal effectively with consumption patterns or to establish sustainable production and consumption strategies has been adopted, and the role of actors who tend to promote unsustainable production and consumption patterns, such as business, is actually strengthened. As well as consumption and production patterns, another distorted linkage to structural economic conditions is made with the recognition that as a result of globalization, external factors have become critical in determining the success or failure of developing countries in their national efforts. It is rightly observed that environmental protection can only be promoted through a shift in the international economy and the establishment of a genuine partnership in order to achieve a more equitable global economy. Yet the idea is that the way to make all countries, in particular developing countries, benefit from globalization is through a combination of trade liberalization, economic development and environmental protection. It is believed that the international trading system should have the capacity to further integrate environmental considerations and enhance its contribution to sustainable development, without undermining its open, equitable and non-discriminatory character.

The text limits itself to recommendations to implement the Uruguay Round and promote trade liberalization. The reality of the present international trading system, a system which promotes discrimination against developing countries, consolidates global disparities and supports unsustainable practices not only in terms of consumption and production but also encouraging transport and pollution and shift from traditional cultures, is not seen as contradictory with the goal of long-term sustainability. With respect to transport, the text notes that the transport sector and mobility in general have an essential and positive role to play in economic and social development, and transportation needs will undoubtedly increase. It also observes that, in the future, transportation is expected to be the major driving force behind a growing world demand for energy. The document accepts that present trends are unsustainable, and adopted recommendations to make transport become more sustainable and mitigate its negative impacts. Yet the document fails to recognize the major cause of transport's expansion, namely, trade liberalization, which encourages production to relocate on the base of a traditional government subsidy to transports or allows for products originating at the other end of the world to be cheaper than products produced a few miles from the consumer. The fact that the whole globalization project is based on the continuity of cheap transport is not discussed.

Generally speaking, UNCED's review was critically received at all levels, being criticized both by diplomats, NGOs and by the press. Ambassador Razali Ismail of Malaysia noted that the compact achieved at Rio had eroded along with much of the high-profile attention to sustainable development generated by UNCED. And the Earth Negotiations Bulletin, a publication of the International Institute for Sustainable Development, noted that "in 1992 one could scarcely escape the news of UNCED and/or the environment in the media. This is not the case today... In international relations, perceptions are everything, and if UNGASS is ultimately billed as a non-event it will not bode well for the future of sustainable development or the UN in general during this critical time of its reform."

Most of the world's press was unanimous in condemning the failure of the New York Summit. The French newspaper Libération, for example, noted in its article "The Earth Summit goes round in circles" that the New York summit closed on an acknowledgment of impotence. Not only did the conference show the little progress accomplished in five years, it also failed to commit governments to significant concrete action and to provide means for implementing Agenda 21. No commitment was taken to achieve the goal of 0.7 % of GDP going to ODA, considered necessary to move towards sustainability. Development assistance today does not exceed 0.3% of GDP, on average, and, in the case of the United States, it was only 0.1 % in 1995.

The US was also the target of much criticism for failing to commit to effectively fighting global warming and to accept concrete reductions in levels of greenhouse gas emissions. At the end of the climate negotiations, no legally binding commitments to target and timetables emerged, and the conference only produced a watery compromise to seek satisfactory results at the then forthcoming Kyoto Conference on Climate Change, which took place in December 1997.

In short, on most major issues at stake, New York 1997 represented a backwards step in relation to UNCED's outcomes. NGOs speak of a scandalous betrayal of the Rio promises and of an utterly shameful outcome from Earth Summit II.

The reality is that the world has changed since Rio, and this change has a name: globalization. The Rio 1992 bargain was based on the commitment by developed countries to provide increased financial resources through ODA and technology transfer to help developing countries move towards sustainability. The implementation of UNCED's agreement was in a sense made dependent upon this aid. However, since Rio, ODA levels have been declining and the private sector has become the major agent of change. Government spending is being cut and state reforms are being carried out worldwide, often reducing not only ODA but also domestic environmental budgets. At UNGASS 1997, developing countries through the G77 tried to obtain a recommitment from the North to UNCED's bargain, including an increase in financial flows, technology transfer and an international economic system more favourable to developing countries. Yet today, as foreign investment replaces overseas development assistance in amount and frequency, UNCED's bargain seems politically outdated, and, as a result, its implementation appears highly jeopardized.

Finally, at the level of NGOs, the fracture among environmentalists is today stronger than five years ago. True, NGOs did lobby the CSD and try to influence the official negotiation process. Indeed, NGOs achieved unprecedented access to the intergovernmental process, with Greenpeace and the Third World Network being allowed to make speeches before the General Assembly. However, most of them had given up the idea of having a unified position on all environmental matters, and no "Global Forum II" was organized in New York, only an inappropriately named "Global Gathering" took place.

CHALLENGES FOR SOCIAL DEVELOPMENT

Some of the important social problems like poverty, ignorance, over-population and rural backwardness are of a general nature and, in varying degree, they are influenced by factors like squalor and bad housing, malnutrition and physical and mental ill-health, neglected childhood, family disorganization and a low standard of living.

For along time, society has remained apathetic to these conditions, but with the awakening of political consciousness and the enthusiasm of organizations and workers to improve social conditions, there is a possibility of developing programmes which could gradually remedy the present situation. The economic programmes of the Five Year Plan will mitigate these problems to some extent, but the gains of economic development have to be maintained and consolidated by well-conceived and organized social welfare programmes spread over the entire country. It is proposed to consider some of the more important problems of social welfare which need the special attention of both State and private welfare agencies.

The principal social welfare problems relate to women, children, youth, the family, under-privileged groups and social service. The social health of any community will depend a great deal upon the status, functions and responsibilities of the woman in the family and in the community. Social conditions should give to the woman opportunities for creative self-expression, so that she can make her full contribution towards the economic and social life of the community. Problems relating to health, maternity and child welfare, education and employment. Some problems of women have to be dealt through social legislation, but other problems pertaining to health, social education, vocational training, and increased participation in social and cultural life, provision of shelter, and assistance to the handicapped or maladjusted call for programmes at the community level. As women have to fulfil heavy domestic and economic responsibilities, adequate attention has to be paid to the need for relaxation and recreation both in the homes as well as in the community. The welfare agencies have catered to some extent to the needs of the widow and the destitute woman, but the quality of the service rendered by them and the nature of their work needs to be surveyed.

Considering the numbers involved, the needs of children should receive much greater consideration than is commonly given to them. There is a growing demand for child health services and educational facilities. The standard of child welfare services in the country can be improved if the rate of increase in population is reduced. Problems relates to family planning, children's health, infant mortality, education, training and development have been discussed elsewhere in this report. Malnutrition is perhaps the major cause of ill-health and lack of proper growth of the child. The feeding of the child in the early years is the responsibility of the family, and is dependent upon economic conditions and traditional food habits. The nature and extent of malnutrition has to be determined, and resources have to be found to supplement and improve the diet of children through schools and community and child welfare agencies.

The problem of children's recreation and development outside educational institutions has received some attention during recent years, but play activities of children are considerably restricted in urban areas on account of the environmental conditions, lack of adequate space, and, to some extent, neglect of this vital need of the child by the family and the community. Not enough is known about the work of private agencies for the welfare of destitute and homeless children. The juvenile courts and children's aid societies have so far touched a fringe of the problem of children's welfare. Certain special aspects may be briefly mentioned. The existing facilities for handicapped and deficient children are far from adequate and suitable agencies have to be created. Hospitals provide treatment for polio, congenital deformities, fractures, bone disorders and other diseases, but there is a need to extend existing services and provide special institutions and care for disabled and crippled children.

At present deficient children attend educational institutions together with normal children and seldom receive treatment and special training to enable them to overcome their handicaps. The subject needs to be studied carefully. The problem of juvenile delinquency has already received considerable attention and many of the States have special legislation. Juvenile delinquency may often be the result of poverty and many offences may be traced to the connivance or support of adults. The youth constitute the most vital section of the community. In recent years, young people have had to face and have been increasingly conscious of problems such as inadequate educational facilities, unemployment, and lack of opportunity for social development, national service and leadership. The problems of health, education and employment of youth have been considered as aspects of national problems in these fields.

Social welfare is primarily concerned with the improvement of services provided for the benefit of youth by welfare agencies with the object of promoting development of character and training for citizenship and for physical, intellectual and moral fitness. It is necessary to encourage initiative among youth so that through their own organizations, they can develop programmes of youth welfare and national service. Ways must also be found to give opportunities to youth for active participation in constructive activity. Such training and experience will equip them for shouldering the responsibilities of leadership in different spheres of national life.

Traditionally, the family has been left largely to its own resources to deal with most of its problems, although in some cases it may be assisted by the larger community groups (such as caste) to which a family may belong. General problems relates to health, education and employment. Questions relating to status and rights, property, inheritance, etc., are the subject of social legislation.

The gradual break-up of the joint family and the emergence of the small family have increased its economic problems and burdens. Family responsibilities have now to be borne at a comparatively younger age by the head of the small family than happened in the joint family. This creates the need for greater guidance and assistance in dealing with family problems. The increasing complexity of the social situation and handicaps arising from physical disability, ailment or unemployment render it more difficult for the family to provide a sense of security to its members. This fact suggests a number of problems which, along with other problems such as divorce, desertion, and treatment of mal-adjusted members of the family, need to be studied carefully if welfare agencies are to develop suitable methods of treatment for guiding and assisting those in need.

There are a number of under-privileged communities such as the scheduled tribes, scheduled castes and other backward classes including criminal tribes. The problems of poverty, ill-health, and lack of opportunities for development affect them to a larger extent than many other sections of the society. The main problems to be considered under the description of social vice are prostitution, crime and delinquency, alcoholism, gambling and beggary. These problems have existed for a long period, although necessarily their nature and extent vary according to the prevailing social and economic conditions. Some of them have to be dealt with largely by local communities, and the approach and treatment have to be varied from place to place.

The character and magnitude of these problems of social defence have to be determined carefully before the value and efficacy of the existing agencies and programmes could be assessed. Social legislation deals with many of the social evils with a view to controlling and even eradicating them, but its actual implementation needs to be watched.

Among the practical problems to be resolved are the demarcations of the relative roles of State and private agencies, determination of the machinery of enforcement, estimation of the resources required, examination of methods, development of correct programmes, and creation of public opinion in favour of an objective and dispassionate approach to the problems of social vice. As the social structure becomes more complex, the State is called upon to play an increasing role in providing services for the welfare of the people. The Central Government, the various State Governments and local self-governing bodies, each in its own sphere, have to ensure that they have at least the minimum administrative machinery for dealing with social "problems. What form this machinery takes will depend on their particular circumstances and requirements, but it is certain that without the necessary machinery they will not be able to pursue their programmes.

Training for Social Work

The contribution which social services make will depend to a considerable extent upon personnel and leadership. A general understanding of the philosophy and history of social work, the structure and functions of society, the nature and extent of social problems, the methods and techniques of social work, and of the details of the programmes and how best their results may be assessed, will help improve the quality and efficacy of all services organized by State and private agencies. The training of social workers should of course include knowledge of conditions prevailing in fields in which they are to work, and social workers must possess the spirit of service and the character and energy to execute programmes despite handicaps and limitations and with such resources-as may be readily available.

There are several schools of social work in India and the setting up of some other institutions on similar lines is being contemplated in some of the States. There are important problems involved in these institutions which require specially qualified and experienced personnel, careful selection of candidates for training, special training for fields in which there is scope for employment, and adequate opportunities for field-work experience. Trained social workers are needed in large numbers for rural areas. It should be possible for the existing schools of social work to draw students from rural areas and to arrange for their training in the field in selected centres organized by rural welfare agencies. Universities and colleges in or near rural areas could also develop training programmes for rural development. Agricultural colleges could introduce intensive social welfare courses and field-work 'programmes as part of their curricula. Similar institutions with greater emphasis on social anthropology could be created in tribal' areas.

It is not possible for many voluntary organizations in the country to employ highly trained personnel for their ordinary programmes and activities. It is, therefore, necessary to arrange for training at the community level for field workers, instructors and supervisors. The existing schools of social work, specialized social service agencies, social welfare agencies functioning at the national and State level should provide opportunities for such training. Arrangements for ' in-service' training should also be made by the larger voluntary organizations which have worked in the field of social welfare for many years. Further, arrangements have to the made for the training of voluntary workers who will be needed in large numbers during the coming years. It is especially desirable that voluntary administrative and field personnel should be given some elementary training in social work.

The emergence of State social services and of large central organizations to deal with important social problems and the lack of opportunities for higher training in the social sciences within the country indicate the need in selected cases for training and study abroad in specialized fields. It is necessary that persons who go abroad for training should first have sufficient knowledge and experience of Indian conditions and problems.

4

Environmental Management and Climate Change

Climate change is any long-term change in the patterns of average weather of a specific region or the Earth as a whole. Climate change reflects abnormal variations to the Earth's climate and subsequent effects on other parts of the Earth, such as in the ice caps over durations ranging from decades to millions of years. In recent usage, especially in the context of environmental policy, climate change usually refers to changes in modern climate. For information on temperature measurements over various periods, and the data sources available.

EFFECTS DUE TO SOLAR VARIATION

Interaction of solar particles, the solar magnetic field, and the Earth's magnetic field, cause variations in the particle and electromagnetic fields at the surface of the planet. Extreme solar events can affect electrical devices. Weakening of the Sun's magnetic field is believed to increase the number of interstellar cosmic rays which reach Earth's atmosphere, altering the types of particles reaching the surface. It has been speculated that a change in cosmic rays could cause an increase in certain types of clouds, affecting Earth's albedo.

Geomagnetic Effects

The Earth's polar aurorae are visual displays created by interactions between the solar wind, the solar magnetosphere, the Earth's magnetic field, and the Earth's atmosphere. Variations in any of these affect aurora displays. Sudden changes can cause the intense disturbances in the Earth's magnetic fields which are called geomagnetic storms.

Solar Proton Events

Energetic protons can reach Earth within 30 minutes of a major flare's peak. During such a solar proton event, Earth is showered in energetic solar particles (primarily protons) released from the flare site. Some of these particles spiral down Earth's magnetic field lines, penetrating the upper layers of our atmosphere where they produce additional ionization and may produce a significant increase in the radiation environment.

Galactic Cosmic Rays

An increase in solar activity (more sunspots) is accompanied by an increase in the "solar wind," which is an outflow of ionized particles, mostly protons and electrons, from the sun. The Earth's geomagnetic field, the solar wind, and the solar magnetic field deflect galactic cosmic rays (GCR). A decrease in solar activity increases the GCR penetration of the troposphere and stratosphere. GCR particles are the primary source of ionization in the troposphere above 1 km (below 1 km, radon is a dominant source of ionization in many areas).

Levels of GCRs have been indirectly recorded by their influence on the production of carbon-14 and beryllium-10. The Hallstatt solar cycle length of approximately 2300 years is reflected by climatic Dansgaard-Oeschger events. The 80–90 year solar Gleissberg cycles appear to vary in length depending upon the lengths of the concurrent 11 year solar cycles, and there also appear to be similar climate patterns occurring on this time scale.

Cloud Effects

Changes in ionization affect the abundance of aerosols that serve as the nuclei of condensation for cloud formation. As a result, ionization levels potentially affect levels of condensation, low clouds, relative humidity, and albedo due to clouds. Clouds formed from greater amounts of condensation nuclei are brighter, longer lived, and likely to produce less precipitation. Changes of 3–4% in cloudiness and concurrent changes in cloud top temperatures have been correlated to the 11 and 22 year solar (sunspot) cycles, with increased GCR levels during "antiparallel" cycles. Global average cloud cover change has been found to be 1.5–2%. Several studies of GCR and cloud cover variations have found positive correlation at latitudes greater than 50° and negative correlation at lower latitudes.

However, not all scientists accept this correlation as statistically significant, and some that do attribute it to other solar variability (e.g. UV or total irradiance variations) rather than directly to GCR changes. Difficulties in interpreting such correlations include the fact that many aspects of solar variability change at similar times, and some climate systems have delayed responses.

Solar Variation Theory

There have been proposals that variations in solar output explain past climate change and contribute to global warming. The most accepted influence of solar variation on the climate is through direct radiative forcing, but this is too small to explain significant temperature change. Various hypotheses have been proposed to explain the apparent solar correlation with temperatures that some assert appear to be stronger than can be explained by direct irradiation and the first order positive feedbacks to increases in solar activity. The meteorological community has responded with skepticism, in part because theories of this nature have come and gone over the course of the 20th century.

Sami Solanki, the director of the Max Planck Institute for Solar System Research in Katlenburg-Lindau, Germany said:

The sun has been at its strongest over the past 60 years and may now be affecting global temperatures... the brighter sun and higher levels of so-called "greenhouse gases" both contributed to the change in the Earth's temperature, but it was impossible to say which had the greater impact.

Nevertheless, Solanki agrees with the scientific consensus that the marked upswing in temperatures since about 1980 is attributable to human activity.

"Just how large this role [of solar variation] is, must still be investigated, since, according to our latest knowledge on the variations of the solar magnetic field, the significant increase in the Earth's temperature since 1980 is indeed to be ascribed to the greenhouse effect caused by carbon dioxide."

The theories have usually represented one of three types:

- Solar irradiance changes directly affecting the climate. This is generally considered unlikely, as the amplitudes of the variations in solar irradiance are much too small to have the observed relation absent some amplification process.
- Variations in the ultraviolet component having an effect. The UV component varies by more than the total, so if UV were for some reason having a disproportionate effect, this might explain a larger solar signal in climate.
- Effects mediated by changes in cosmic rays (which are affected by the solar wind, which is affected by the solar output) such as changes in cloud cover.

Although correlations often can be found, the mechanism behind these correlations is a matter of speculation. Many of these speculative accounts have fared badly over time, and in a paper "Solar activity and terrestrial climate: an analysis of some purported correlations." Peter Laut demonstrates problems with some of the most popular, notably those by Svensmark and by Lassen. Damon and Laut report in Eos that *the apparent strong correlations displayed on these graphs have been obtained by incorrect handling of the physical data. The graphs are still widely referred to in the literature, and their misleading character has not yet been generally recognized.*

In 1991, Knud Lassen of the Danish Meteorological Institute in Copenhagen and his colleague Eigil Friis-Christensen found a strong correlation between the length of the solar cycle and temperature changes throughout the northern hemisphere. Initially, they used sunspot and temperature measurements from 1861 to 1989, but later found that climate records dating back four centuries supported their findings. This relationship appeared to account for nearly 80 per cent of the measured temperature changes over this period.

Damon and Laut, however, show that when the graphs are corrected for filtering errors, *the sensational agreement with the recent global warming, which drew worldwide attention, has totally disappeared. Nevertheless, the authors and other researchers keep presenting the old misleading graph*. Note that the prior link to "graph" is one such example of this.

Sallie Baliunas, an astronomer at the Harvard-Smithsonian Centre for Astrophysics, has been among the supporters of the theory that changes in the sun "can account for major climate changes on Earth for the past 300 years, including part of the recent surge of global warming."

On May 6, 2000, however, *New Scientist* magazine reported that Lassen and astrophysicist Peter Thejll had updated Lassen's 1991 research and found that while the solar cycle still accounts for about half the temperature rise since 1900, it fails to explain a rise of 0.4 °C since 1980. "The curves diverge after 1980," Thejll said, "and it's a startlingly large deviation. Something else is acting on the climate.... It has the fingerprints of the greenhouse effect."

Later that same year, Peter Stott and other researchers at the Hadley Centre in the United Kingdom published a paper in which they reported on the most comprehensive model simulations to date of the climate of the 20th century.

Their study looked at both "natural forcing agents" (solar variations and volcanic emissions) as well as "anthropogenic forcing" (greenhouse gases and sulphate aerosols). They found that "solar effects may have contributed significantly to the warming in the first half of the century although this result is dependent on the reconstruction of total solar irradiance that is used.

In the latter half of the century, we find that anthropogenic increases in greenhouses gases are largely responsible for the observed warming, balanced by some cooling due to anthropogenic sulphate aerosols, with no evidence for significant solar effects." Stott's team found that combining all of these factors enabled them to closely simulate global temperature changes throughout the 20th century. They predicted that continued greenhouse gas emissions would cause additional future temperature increases "at a rate similar to that observed in recent decades".

It should be noted that their solar forcing included "spectrally-resolved changes in solar irradiance" and not the indirect effects mediated through cosmic rays for which there is still no accepted mechanism — these ideas are still being fleshed out. In addition, the study notes "uncertainties in historical forcing" — in other words, past natural forcing may still be having a delayed warming effect, most likely due to the oceans.

A graphical representation of the relationship between natural and anthropogenic factors contributing to climate change appears in "Climate Change 2001: The Scientific Basis", a report by the Intergovernmental Panel on Climate Change (IPCC). Stott's 2003 work mentioned in the model section above largely revised his assessment, and found a significant solar contribution to recent warming, although still smaller (between 16 and 36%) than that of the greenhouse gases.

Physicist and historian Spencer R. Weart in *The Discovery of Global Warming* (2003) writes:

The study of [sun spot] cycles was generally popular through the first half of the century. Governments had collected a lot of weather data to play with and inevitably people found correlations between sun spot cycles and select weather patterns. If rainfall in England didn't fit the cycle, maybe storminess in New England would. Respected scientists and enthusiastic amateurs insisted they had found patterns reliable enough to make predictions. Sooner or later though every prediction failed.

An example was a highly credible forecast of a dry spell in Africa during the sunspot minimum of the early 1930s. When the period turned out to be wet, a meteorologist later recalled "the subject of sunspots and weather relationships fell into dispute, especially among British meteorologists who witnessed the discomfiture of some of their most respected superiors." Even in the 1960s he said, "For a young [climate] researcher to entertain any statement of sun-weather relationships was to brand oneself a crank.")

Orbital Variations

In their effect on climate, orbital variations are in some sense an extension of solar variability, because slight variations in the Earth's orbit lead to changes in the distribution and abundance of sunlight reaching the Earth's surface. These orbital variations, known as Milankovitch cycles, directly affect glacial activity. Eccentricity, axial tilt, and precession comprise the three dominant cycles that make up the variations in Earth's orbit.

The combined effect of the variations in these three cycles creates changes in the seasonal reception of solar radiation on the Earth's surface. As such,

Milankovitch Cycles affecting the increase or decrease of received solar radiation directly influence the Earth's climate system, and influence the advance and retreat of Earth's glaciers. Subtler variations are also present, such as the repeated advance and retreat of the Sahara desert in response to orbital precession.

Volcanism

Volcanism is the process of conveying material from the depths of the Earth to the surface, as part of the process by which the planet removes excess heat and pressure from its interior. Volcanic eruptions, geysers and hot springs are all part of the volcanic process and all release varying levels of particulates into the atmosphere.

A single eruption of the kind that occurs several times per century can affect climate, causing cooling for a period of a few years or more. The eruption of Mount Pinatubo in 1991, for example, produced the second largest terrestrial eruption of the 20th century (after the 1912 eruption of Novarupta)and affected the climate substantially, with global temperatures dropping by about 0.5 °C (0.9 °F), and ozone depletion being temporarily substantially increased.

Much larger eruptions, known as large igneous provinces, occur only a few times every hundred million years, but can reshape climate for millions of years and cause mass extinctions. Initially, it was thought that the dust ejected into the atmosphere from large volcanic eruptions was responsible for longer-term cooling by partially blocking the transmission of solar radiation to the Earth's surface. However, measurements indicate that most of the dust hurled into the atmosphere may return to the Earth's surface within as little as six months, given the right conditions.

Volcanoes are also part of the extended carbon cycle. Over very long (geological) time periods, they release carbon dioxide from the Earth's interior, counteracting the uptake by sedimentary rocks and other geological carbon dioxide sinks. According to the US Geological Survey, however, estimates are that human activities generate more than 130 times the amount of carbon dioxide emitted by volcanoes.

Ocean Variability

On a timescale often measured in decades or more, climate changes can also result from the interaction between the atmosphere and the oceans. Many climate fluctuations, including the El Niño Southern oscillation, the Pacific decadal oscillation, the North Atlantic oscillation, and the Arctic oscillation, owe their existence at least in part to the different ways that heat may be stored in the oceans and also to the way it moves between various 'reservoirs'.

On longer time scales (with a complete cycle often taking up to a thousand years to complete), ocean processes such as thermohaline circulation also play a key role in redistributing heat by carrying out a very slow and extremely deep movement of water, and the long-term redistribution of heat in the oceans.

The term thermohaline circulation (THC) refers to the part of the large-scale ocean circulation that is driven by global density gradients created by surface heat and freshwater fluxes. The adjective thermohaline derives from *thermo-* referring to temperature and *-haline* referring to salt content, factors which together determine the density of sea water. Wind-driven surface currents (such as the Gulf Stream) head polewards from the equatorial Atlantic Ocean, cooling all the while and eventually sinking at high latitudes (forming North Atlantic Deep Water). This dense water then flows into the ocean basins.

While the bulk of it upwells in the Southern Ocean, the oldest waters (with a transit time of around 1600 years) upwell in the North Pacific. Extensive mixing therefore takes place between the ocean basins, reducing differences between them and making the Earth's ocean a global system. On their journey, the water masses transport both energy (in the form of heat) and matter (solids, dissolved substances and gases) around the globe. As such, the state of the circulation has a large impact on the climate of the Earth.

The thermohaline circulation is sometimes called the ocean conveyor belt, the great ocean conveyor, or the global conveyor belt. On occasion, it is used to refer to the meridional overturning circulation (often abbreviated as MOC). The term MOC, however, is more accurate and well defined, as it is difficult to separate the part of the circulation which is actually driven by temperature and salinity alone as opposed to other factors such as the wind. Temperature and salinity gradients can also lead to a circulation which does not add to the MOC itself.

The movement of surface currents pushed by the wind is intuitive: we have all seen wind ripples on the surface of a pond. Thus the deep ocean — devoid of wind — was assumed to be perfectly static by early oceanographers. However, modern instrumentation shows that current velocities in deep water masses can be significant (although much less than surface speeds). In the deep ocean, the predominant driving force is differences in density, caused by salinity and temperature (the more saline the denser, and the colder the denser).

There is often confusion over the components of the circulation that are wind and density driven. Note that ocean currents due to tides are also significant in many places; most prominent in relatively shallow coastal areas, tidal currents can also be significant in the deep ocean. The density of ocean water is not globally homogeneous, but varies significantly and discretely. Sharply defined boundaries exist between water masses which form at the surface, and subsequently maintain their own identity within the ocean.

They position themselves one above or below each other according to their density, which depends on both temperature and salinity. Warm seawater expands and is thus less dense than cooler seawater. Saltier water is denser than fresher water because the dissolved salts fill interstices between water molecules, resulting in more mass per unit volume. Lighter water masses float over denser ones (just as a piece of wood or ice will float on water).

This is known as "stable stratification". When dense water masses are first formed, they are not stably stratified. In order to take up their most stable positions, water masses of different densities must flow, providing a driving force for deep currents. The thermohaline circulation is mainly triggered by the formation of deep water masses in the North Atlantic and the Southern Ocean and Haline forcing caused by differences in temperature and salinity of the water.

Formation of Deep Water Masses

The dense water masses that sink into the deep basins are formed in quite specific areas of the North Atlantic and the Southern Ocean. In these Polar Regions, seawater at the surface of the ocean is intensively cooled by the wind. Wind moving over the water also produces a great deal of evaporation, leading to a decrease in temperature, called evaporative cooling. Evaporation removes only molecules of pure water, resulting in an increase in the salinity of the seawater left behind, and thus an increase in the density of the water mass.

In the Norwegian Sea evaporative cooling is predominant, and the sinking water mass, the North Atlantic Deep Water (NADW), fills the basin and spills southwards through crevasses in the submarine sills that connect Greenland, Iceland and Great Britain. It then flows very slowly into the deep abyssal plains of the Atlantic, always in a southerly direction. Flow from the Arctic Ocean Basin into the Pacific, however, is blocked by the narrow shallows of the Bering Strait.

The formation of sea ice also contributes to an increase in seawater salinity; saltier brine is left behind as the sea ice forms around it (pure water preferentially being frozen). Increasing salinity depresses the freezing temperature of seawater, so cold liquid brine is formed in inclusions within a honeycomb of ice. The brine progressively melts the ice just beneath it, eventually dripping out of the ice matrix and sinking.

This process is known as brine exclusion. By contrast in the Weddell Sea off the coast of Antarctica near the edge of the ice pack, the effect of wind cooling is intensified by brine exclusion. The resulting Antarctic Bottom Water

(AABW) sinks and flows north into the Atlantic Basin, but is so dense it actually underflows the NADW. Again, flow into the Pacific is blocked, this time by the Drake Passage between the Antarctic Peninsula and the southernmost tip of South America.

The dense water masses formed by these processes flow downhill at the bottom of the ocean, like a stream within the surrounding less dense fluid, and fill up the basins of the polar seas. Just as river valleys direct streams and rivers on the continents, the bottom topography steers the deep and bottom water masses. Note that, unlike fresh water, saline water does not have a density maximum at 4°C but gets denser as it cools all the way to its freezing point of approximately "1.8°C.

Movement of Thermohaline Circulation

Formation and movement of the deep water masses at the North Atlantic Ocean, creates sinking water masses that fill the basin and flows very slowly into the deep abyssal plains of the Atlantic. This high latitude cooling and the low latitude heating drives the movement of the deep water in a polar southward flow.

The deep water flows through the Antarctic Ocean Basin around South Africa where it is split into two routes: one into the Indian Ocean and one past Australia into the Pacific.

At the Indian Ocean, some of the cold and salty water from Atlantic — drawn by the flow of warmer and fresher upper ocean water from the tropical Pacific — causes a vertical exchange of dense, sinking water with lighter water above. It is known as overturning. In the Pacific Ocean, the rest of the cold and salty water from the Atlantic undergoes Haline forcing and slowly becomes warmer and fresher.

The out-flowing undersea of cold and salty water makes the sea level of the Atlantic slightly lower than the Pacific and salinity or halinity of water at the Atlantic higher than the Pacific. This generates a large but slow flow of warmer and fresher upper ocean water from the tropical Pacific to the Indian Ocean through the Indonesian Archipelago to replace the cold and salty Antarctic Bottom Water. This is also known as Haline forcing (net high latitude freshwater gain and low latitude evaporation).

This warmer, fresher water from the Pacific flows up through the South Atlantic to Greenland, where it cools off and undergoes evaporative cooling and sinks to the ocean floor, providing a continuous thermohaline circulation. Hence, a recent and popular name for the thermohaline circulation, emphasizing the vertical nature and pole-to-pole character of this kind of ocean circulation, is the meridional overturning circulation.

Quantitative Estimation

The deep water masses that participate in the MOC have chemical, temperature and isotopic ratio signatures and can be traced, their flow rate calculated, and their age determined. These include ^{231}Pa / ^{230}Th ratios.

Gulf Stream

The North Atlantic Current, warm ocean current that continues the Gulf Stream northeast, is largely driven by the global thermohaline circulation to further east and north from the North American coast, across the Atlantic and into the Arctic Ocean.

Upwelling

All these dense water masses sinking into the ocean basins displace the water below them, so that elsewhere water must be rising in order to maintain a balance. However, because this thermohaline upwelling is so widespread and diffuse, its speeds are very slow even compared to the movement of the bottom water masses.

It is therefore difficult to measure where upwelling occurs using current speeds, given all the other wind-driven processes going on in the surface ocean.

Deep waters do however have their own chemical signature, formed from the breakdown of particulate matter falling into them over the course of their long journey at depth; and a number of authors have tried to use these tracers to infer where the upwelling occurs. Wallace Broecker, using box models, has asserted that the bulk of deep upwelling occurs in the North Pacific, using as evidence the high values of silicon found in these waters. However, other investigators have not found such clear evidence.

Computer models of ocean circulation increasingly place most of the deep upwelling in the Southern Ocean, associated with the strong winds in the open latitudes between South America and Antarctica. While this picture is consistent with the global observational synthesis of William Schmitz at Woods Hole and with low observed values of diffusion, not all observational syntheses agree. Recent papers by Lynne Talley at the Scripps Institution of Oceanography and Bernadette Sloyan and Stephen Rintoul in Australia suggest that a significant amount of dense deep water must be transformed to light water somewhere north of the Southern Ocean.

Effects on Global Climate

The thermohaline circulation plays an important role in supplying heat to the polar regions, and thus in regulating the amount of sea ice in these regions. Changes in the thermohaline circulation are thought to have significant impacts on the earth's radiation budget. Insofar as the thermohaline circulation governs the rate at which deep waters are exposed to the surface, it may also play an important role in determining the concentration of carbon dioxide in the atmosphere.

While it is often stated that the thermohaline circulation is the primary reason that Western Europe is so temperate, it has been suggested that this is largely incorrect, and that Europe is warm mostly because it lies downwind of an ocean basin, and because of the effect of atmospheric waves bringing warm air north from the subtropics. However, the underlying assumptions of this particular analysis have been challenged.

Large influxes of low density meltwater from Lake Agassiz and deglaciation in North America are thought to have led to a disruption of deep water formation and subsidence in the extreme North Atlantic and caused the climate period in Europe known as the Younger Dryas. For a discussion of the possibilities of changes to the thermohaline circulation under global warming.

Human Influences

Anthropogenic factors are human activities that change the environment. In some cases the chain of causality of human influence on the climate is direct and unambiguous (for example, the effects of irrigation on local humidity), whilst in other instances it is less clear. Various hypotheses for human-induced climate change have been argued for many years though, generally, the scientific debate has moved on from scepticism to a scientific consensus on climate change that human activity is the probable cause for the rapid changes in world climate in the past several decades.

Consequently, the debate has largely shifted onto ways to reduce further human impact and to find ways to adapt to change that has already occurred. Of most concern in these anthropogenic factors is the increase in CO_2 levels due to emissions from fossil fuel combustion, followed by aerosols (particulate matter in the atmosphere) and cement manufacture. Other factors, including land use, ozone depletion, animal agriculture and deforestation, are also of concern in the roles they play - both separately and in conjunction with other factors - in affecting climate.

INTERGOVERNMENTAL PANEL ON CLIMATE CHANGE

The sum of these components indicates a rate of eustatic sea level rise (corresponding to a change in ocean volume) from 1910 to 1990 ranging from –0.8 to 2.2 mm/yr, with a central value of 0.7 mm/yr. The upper bound is close to the observational upper bound (2.0 mm/yr), but the central value is less than the observational lower bound (1.0 mm/yr), i.e., the sum of components is biased low compared to the observational estimates.

The sum of components indicates an acceleration of only 0.2 (mm/yr)/ century, with a range from –1.1 to +0.7 (mm/yr)/century, consistent with observational finding of no acceleration in sea level rise during the 20th century. The estimated rate of sea-level rise from anthropogenic climate change from 1910 to 1990 (from modeling studies of thermal expansion, glaciers and ice sheets) ranges from 0.3 to 0.8 mm/yr. It is very likely that 20th century warming has contributed significantly to the observed sea-level rise, through thermal expansion of sea water and widespread loss of land ice.

A common perception is that the rate of sea-level rise should have accelerated during the latter half of the 20th century, but tide gauge data for the 20th century show no significant acceleration. Estimates obtained are based on AOGCMs for the terms directly related to anthropogenic climate change in the 20th century, i.e., thermal expansion, ice sheets, glaciers and ice caps... The total computed rise indicates an acceleration of only 0.2 (mm/yr)/ century, with a range from -1.1 to +0.7 (mm/yr)/century, consistent with observational finding of no acceleration in sea-level rise during the 20th century.

The sum of terms not related to recent climate change is -1.1 to +0.9 mm/ yr (i.e., excluding thermal expansion, glaciers and ice caps, and changes in the ice sheets due to 20th century climate change). This range is less than the observational lower bound of sea level rise.

Hence it is very likely that these terms alone are an insufficient explanation, implying that 20th century climate change has made a contribution to 20th century sea level rise. Recent figures of human, terrestrial impoundment came too late for the 3rd Report, and would revise levels upward for much of the 20th century.

Uncertainties and criticisms regarding IPCC results:

- Tide records with a rate of 180 mm/century going back to the 19th century show no measurable acceleration throughout the late 19th and first half of the 20th century. The IPCC attributes about 60 mm/century to melting and other eustatic processes, leaving a residual of 120 mm of 20th century rise to be accounted for. Global ocean temperatures by Levitus et al. are in accord with coupled ocean/atmosphere modelling of greenhouse warming, with heat-related change of 30 mm. Melting of polar ice sheets at the upper limit of the IPCC estimates could close the gap, but severe limits are imposed by the observed perturbations in Earth rotation.
- By the time of the IPCC TAR, attribution of sea-level changes had a large unexplained gap between direct and indirect estimates of global

sea-level rise. Most direct estimates from tide gauges give 1.5–2.0 mm/yr, whereas indirect estimates based on the two processes responsible for global sea-level rise, namely mass and volume change, are significantly below this range. Estimates of the volume increase due to ocean warming give a rate of about 0.5 mm/yr and the rate due to mass increase, primarily from the melting of continental ice, is thought to be even smaller. One study confirmed tide gauge data is correct, and concluded there must be a continental source of 1.4 mm/yr of fresh water.

- From: "In the last dozen years, published values of 20th century GSL rise have ranged from 1.0 to 2.4 mm/yr. In its Third Assessment Report, the IPCC discusses this lack of consensus at length and is careful not to present a best estimate of 20th century GSL rise. By design, the panel presents a snapshot of published analysis over the previous decade or so and interprets the broad range of estimates as reflecting the uncertainty of our knowledge of GSL rise. We disagree with the IPCC interpretation. In our view, values much below 2 mm/yr are inconsistent with regional observations of sea-level rise and with the continuing physical response of Earth to the most recent episode of deglaciation."
- The strong 1997-1998 El Niño caused regional and global sea level variations, including a temporary global increase of perhaps 20 mm. The IPCC TAR's examination of satellite trends says *the major 1997/98 El Niño-Southern Oscillation (ENSO) event could bias the above estimates of sea-level rise and also indicate the difficulty of separating long-term trends from climatic variability.*

Glacier Contribution

It is well known that glaciers are subject to surges in their rate of movement with consequent melting when they reach lower altitudes and/or the sea. Historical reports of surge occurrences in Iceland's glaciers go back several centuries. Thus rapid retreat can have several other causes than CO_2 increase in the atmosphere. The results from Dyurgerov show a sharp increase in the contribution of mountain and subpolar glaciers to sea level rise since 1996 (0.5 mm/yr) to 1998 (2 mm/yr) with an average of approx. 0.35 mm/yr since 1960. Of interest also is Arendt et al., who estimate the contribution of Alaskan glaciers of 0.14±0.04 mm/yr between the mid 1950s to the mid 1990s increasing to 0.27 mm/yr in the middle and late 1990s.

Greenland Contribution

Krabill *et al.* estimate a net contribution from Greenland to be at least 0.13 mm/yr in the 1990s. Joughin *et al.* have measured a doubling of the speed of Jakobshavn Isbræ between 1997 and 2003. This is Greenland's largest-outlet glacier; it drains 6.5% of the ice sheet, and is thought to be responsible for increasing the rate of sea level rise by about 0.06 millimeters per year, or roughly 4% of the 20th century rate of sea level increase. In 2004, Rignot *et al.* estimated a contribution of 0.04±0.01 mm/yr to sea level rise from southeast Greenland.

Rignot and Kanagaratnam produced a comprehensive study and map of the outlet glaciers and basins of Greenland. They found widespread glacial acceleration below 66 N in 1996 which spread to 70 N by 2005; and that the ice sheet loss rate in that decade increased from 90 to 200 cubic km/yr; this corresponds to an extra 0.25 to 0.55 mm/yr of sea level rise.

In July 2005 it was reported that the Kangerdlugssuaq glacier, on Greenland's east coast, was moving towards the sea three times faster than a decade earlier. Kangerdlugssuaq is around 1,000 m thick, 7.2 km (4.5 miles) wide, and drains about 4% of the ice from the Greenland ice sheet. Measurements of Kangerdlugssuaq in 1988 and 1996 showed it moving at between 5 and 6 km/yr (3.1 to 3.7 miles/yr) (in 2005 it was moving at 14 km/yr [8.7 miles/yr]).

According to the 2004 Arctic Climate Impact Assessment, climate models project that local warming in Greenland will exceed 3° Celsius during this century. Also, ice sheet models project that such a warming would initiate the long-term melting of the ice sheet, leading to a complete melting of the Greenland ice sheet over several millennia, resulting in a global sea level rise of about seven meters.

Antarctic Contribution

On the Antarctic continent itself, the large volume of ice present stores around 70 % of the world's fresh water. This ice sheet is constantly gaining ice from snowfall and losing ice through outflow to the sea. West Antarctica is currently experiencing a net outflow of glacial ice, which will increase global sea level over time. A review of the scientific studies looking at data from 1992 to 2006 suggested a net loss of around 50 Gigatonnes of ice per year was a reasonable estimate (around 0.14 mm of sea level rise).

Although, significant acceleration of outflow glaciers in the Amundsen Sea Embayment could have more than doubled this figure for the year 2006. East Antarctica is a cold region with a ground base above sea level and occupies most of the continent. This area is dominated by small accumulations of snowfall which becomes ice and thus eventually seaward glacial flows. The mass balance of the East Antarctic Ice Sheet as a whole is thought to be slightly positive (lowering sea level) or near to balance. However, increased ice outflow has been suggested in some regions.

Effects of Snowline and Permafrost

The snowline altitude is the altitude of the lowest elevation interval in which minimum annual snow cover exceeds 50%. This ranges from about 5,500 metres above sea-level at the equator down to sea-level at about 65° N&S latitude, depending on regional temperature amelioration effects. Permafrost then appears at sea-level and extends deeper below sea-level pole-wards. The depth of permafrost and the height of the ice-fields in both Greenland and Antarctica means that they are largely invulnerable to rapid melting. Greenland Summit is at 3,200 metres, where the average annual temperature is minus 32 °C. So even a projected 4 °C rise in temperature leaves it well below the melting point of ice.

Frozen Ground 28, December 2004, has a very significant map of permafrost affected areas in the Arctic. The continuous permafrost zone includes all of Greenland, the North of Labrador, NW Territories, Alaska north of Fairbanks, and most of NE Siberia north of Mongolia and Kamchatka. Continental ice above permafrost is very unlikely to melt quickly. As most of the Greenland and Antarctic ice sheets lie above the snowline and/or base of the permafrost zone, they cannot melt in a timeframe much less than several millennia; therefore they are unlikely to contribute significantly to sea-level rise in the coming century.

Polar Ice

The sea level will rise above its current level if more polar ice melts. However, compared to the heights of the ice ages, today there are very few continental ice sheets remaining to be melted. It is estimated that Antarctica, if fully melted, would contribute more than 60 metres of sea level rise, and Greenland would contribute more than 7 metres. Small glaciers and ice caps on the margins of Greenland and the Antarctic Peninsula might contribute about 0.5 metres.

While the latter figure is much smaller than for Antarctica or Greenland it could occur relatively quickly (within the coming century) whereas melting of Greenland would be slow (perhaps 1,500 years to fully deglaciate at the fastest likely rate) and Antarctica even slower. However, this calculation does not account for the possibility that as meltwater flows under and lubricates the larger ice sheets, they could begin to move much more rapidly towards the sea.

In 2002, Rignot and Thomas found that the West Antarctic and Greenland ice sheets were losing mass, while the East Antarctic ice sheet was probably in balance (although they could not determine the sign of the mass balance for The East Antarctic ice sheet). Kwok and Comiso also discovered that temperature and pressure anomalies around West Antarctica and on the other side of the Antarctic Peninsula correlate with recent Southern Oscillation events.

In 2004 Rignot et al. estimated a contribution of 0.04±0.01 mm/yr to sea level rise from South East Greenland. In the same year, Thomas et al. found evidence of an accelerated contribution to sea level rise from West Antarctica.

The data showed that the Amundsen Sea sector of the West Antarctic Ice Sheet was discharging 250 cubic kilometres of ice every year, which was 60% more than precipitation accumulation in the catchment areas.

This alone was sufficient to raise sea level at 0.24 mm/yr. Further, thinning rates for the glaciers studied in 2002-2003 had increased over the values measured in the early 1990s. The bedrock underlying the glaciers was found to be hundreds of meters deeper than previously known, indicating exit routes for ice from further inland in the Byrd Subpolar Basin.

Thus the West Antarctic ice sheet may not be as stable as has been supposed. In 2005 it was reported that during 1992-2003, East Antarctica thickened at an average rate of about 18 mm/yr while West Antarctica showed an overall thinning of 9 mm/yr. associated with increased precipitation. A gain of this magnitude is enough to slow sea-level rise by 0.12±0.02 mm/yr.

Effects of Sea Level Rise

Based on the projected increases stated above, the IPCC TAR WG II report notes that current and future climate change would be expected to have a number of impacts, particularly on coastal systems. Such impacts may include increased coastal erosion, higher storm-surge flooding, inhibition of primary production processes, more extensive coastal inundation, changes in surface water quality and groundwater characteristics, increased loss of property and coastal habitats, increased flood risk and potential loss of life, loss of nonmonetary cultural resources and values, impacts on agriculture and aquaculture through decline in soil and water quality, and loss of tourism, recreation, and transportation functions.

There is an implication that many of these impacts will be detrimental—especially for the three-quarters of the world's poor who depend on agriculture systems. The report does, however, note that owing to the great diversity of coastal environments; regional and local differences in projected relative sea level and climate changes; and differences in the resilience and adaptive capacity of ecosystems, sectors, and countries, the impacts will be highly variable in time and space.

Statistical data on the human impact of sea level rise is scarce. A study in the April, 2007 issue of *Environment and Urbanization* reports that 634 million people live in coastal areas within 30 feet (9.1 m) of sea level. The study also reported that about two thirds of the world's cities with over five million people are located in these low-lying coastal areas.

The IPCC report of 2007 estimated that accelerated melting of the Himalayan ice caps and the resulting rise in sea levels would likely increase the severity of flooding in the short-term during the rainy season and greatly magnify the impact of tidal storm surges during the cyclone season. A sea-level rise of just 40 cm in the Bay of Bengal would put 11 per cent of the country's coastal land underwater, creating 7 to 10 million climate refugees.

Island Nations

IPCC assessments suggest that deltas and small island states are particularly vulnerable to sea level rise caused by both thermal expansion and ocean volume. Relative sea level rise (mostly caused by subsidence) is currently causing substantial loss of lands in some deltas.

Sea level changes have not yet been conclusively proven to have directly resulted in environmental, humanitarian, or economic losses to small island states, but the IPCC and other bodies have found this a serious risk scenario in coming decades.

Many media reports have focused the island nations of the Pacific, notably the Polynesian islands of Tuvalu, which based on more severe flooding events in recent years, was thought to be "sinking" due to sea level rise. A scientific review in 2000 reported that based on University of Hawaii gauge data, Tuvalu had experienced a negligible increase in sea-level of 0.07 mm a year over the past two decades, and that ENSO had been a larger factor in Tuvalu's higher tides in recent years.

A subsequent study by John Hunter from the University of Tasmania, however, adjusted for ENSO effects and the movement of the gauge (which was thought to be sinking).

Hunter concluded that Tuvalu had been experiencing sea-level rise of about 1.2 mm per year. The recent more frequent flooding in Tuvalu may also be due to an erosional loss of land during and following the actions of 1997 cyclones Gavin, Hina, and Keli.

Reuters has reported other Pacific islands are facing a severe risk including Tegua island in Vanuatu. Claims that Vanuatu data shows no net sea level rise, are not substantiated by tide gauge data. Vanuatu tide gauge data show a net rise of ~50 mm from 1994-2004. Linear regression of this short time series suggests a rate of rise of ~7 mm/y, though there is considerable variability and the exact threat to the islands is difficult to assess using such a short time series. Numerous options have been proposed that would assist island nations to adapt to rising sea level.

THE CHALLENGE OF ENVIRONMENTAL MANAGEMENT

Before moving on to introduce the elements of management practice and strategy which will facilitate improved environmental performance in industry, it is important to relay some of the factors which together serve to encourage industry to respond to the environmental challenge.

Environmental Efficiency

Companies often strive to minimize the costs of their operations. This is especially relevant in relation to the efficiency with which they use their material inputs. As the ability of the environment to supply raw materials and accept waste is diminished, the costs of these services to industry will increase. As a result, more efficient raw material utilization and a decrease in the amount of waste generated are key factors which will encourage industry to minimize its environmental impact. Particularly in relation to waste products, companies are experiencing increasingly stringent legislation which increases the costs of waste management. Waste should be viewed both in terms of physical waste generated and the less tangible losses experienced through an inefficient use of resources. Avoiding these losses improves both the business and environmental performance of a company. As a result, many companies have pursued a strategy of waste minimization for a number of years and have experienced short payback periods on investment in waste management. In efforts to increase the efficiency of their operations, many companies have developed integrated management systems to reduce inefficiencies and the likelihood of errors. Most commonly to date, these have centered around the promotion of quality.

The Influence of Government

The main impact of government on the environmental performance of industry has been through the development of environmental legislation. Environmental considerations have been built into the legislative framework for many years. Initially, establishing rights of ownership over natural resources led to the development of a legal system to protect those rights. Subsequently, the impact of industrial activity on the health of employees and the surrounding community *kd* to the creation of public health and safety legislation. Measures have also been introduced to control the use of products, processes and wastes which may harm the environment. The impact of environmental legislation on the operation of industry has been profound and is set to become ever more tough. As the strain placed upon the environment mounts and knowledge of the causes and effects of environmental degradation becomes more complete, the extent and impact of environmental legislation will continue to develop.

Thus, industry must satisfy an increasing number of legal obligations in relation to the effect that its activities have upon the environment. As a result, in all of its operations, industry must plan ahead to meet the demands of current and forthcoming environmental legislation. By developing proactive responses to legislative pressure, industry will reduce its costs and exposure to risk.

While in the short term, legal obligations undoubtedly increase the costs of production that fall upon the firm, it is up to each firm to comply with legislation in the most cost-effective way. The development of proactive strategic responses to the demands of legislation will reduce these costs. In parallel with the development of environmental legislation, governments are increasingly applying market instruments to achieve environmental objectives. Actions of this nature may include the imposition of taxes on environmentally damaging goods, subsidies on environmentally friendly goods or the provision of information relating to the environmental performance of companies or products. Market instruments are intended to channel the choice of consumers or other stakeholders towards the better environmental option. Thus through a combination of legislative and market instruments, by encouraging certain activities and discouraging others, governments seek to accelerate the structural change which encourages improved environmental efficiency in the economy as a whole.

The Development of Stakeholder Influence

Individual businesses interact with a number of stakeholders, all of whom have an interest in the performance of that company. Traditionally the main focus of stakeholder interest has been upon the financial performance of the company. Increasingly, however, stakeholder pressure is concentrating on the environmental performance of the company. The range of stakeholders which demand high environmental standards.

Customers

The relationship between a company and its customers is obviously of para-mount importance. In relation to environmental considerations, the potential importance of green consumerism cannot be overstated. The range of characteristics that underlay the purchasing decision are a fundamental consideration for all businesses. Increasingly, the environment is being accepted as one such characteristic by consumers. At present, however, the influence of green consumerism on most businesses is marginal. Of the myriad of products that each consumer buys, very few are chosen on the basis of their environmental credentials alone. Nevertheless, it is certain that credible claims relating to environmental performance constitute one positive element among the many characteristics upon which consumers base their purchasing decision. Companies which can validate and communicate the environmental performance of their products will enhance their competitive position. Governments are also seeking to increase the potency of green consumerism by providing the consumer with the information necessary to make an informed choice in relation to the environmental performance of each product within the product range. For this reason for example, we saw the introduction of the EC's eco-labelling scheme in 1993.

Trading Partners

Many businesses do not sell into 'end-consumer' markets and may therefore perceive themselves to be remote from any consumer pressures to improve their environmental performance. Increasingly, the pressure to improve environmental performance is emanating from trading partners rather than the ultimate consumer. In efforts to improve overall environmental performance, many companies are exercising their own rights both as purchasers and as vendors and are demanding that all of the companies within their supply chain seek to minimize their own environmental impacts. Hence, demands to improve environmental performance at all stages in the supply chain are being diffused beyond those companies that are directly exposed to the pressures of green consumerism. An increasing number of companies are preferring to buy their resources from or sell their products to companies which meet certain standards of environmental performance.

The provision of information on company environmental performance through standards such as BS7750, the British Standard on Environmental Management Systems, and the EC's eco-management and audit scheme will increasingly be written into contracts in the future. Increasing environmental concern and the improved provision of information relating to environmental performance will reward those companies which achieve and communicate high environmental standards with a competitive advantage.

The Community

Industry shares its surrounding environment with the local population. Increasingly this population is demanding a high level of environmental performance from its industrial neighbours, and seeks some degree of reassurance that they are not exposed to significant environmental risk due to a company's operations. This concern has been recognized for many years and was initially recognized in public health legislation. Trends towards freedom of access to environmental information will give greater power to local communities when they question the activities of local industrial co-habitants. In order to foster a positive working relationship, companies must improve their environmental performance and communicate their efforts to the surrounding communities. This is true both for future developments and existing operations.

Employees

The population in the community surrounding a company also includes the workforce of that company. The pressure to provide a healthy living environment is magnified within the workplace. Employees seek healthy and secure working conditions, and can draw on an established framework of health and safety legislation in this respect. However, employees' concerns relating to the environmental performance of their employers goes beyond the impact of operations on the working and living environment. Increasingly people wish to work for ethical and responsible companies. Companies that reflect the environmental concerns of the public will find it easier to attract, retain and motivate a quality workforce.

Investors and Insurers

The pressures to improve environmental performance also emanate from the investors and shareholders of a company. The rapid growth of ethical investment schemes in recent years reflects the desire of many investors only to lend their financial support to companies which behave in a responsible manner. There are also a number of very good business reasons why investors prefer to work with companies that have a proven track record of environmental integrity. The structure of legal liability for environmental damage dictates that any party that causes environmental damage may be fined and required to bear the costs of remediating that damage and to compensate the affected parties for any associated losses. It is increasingly difficult and expensive to obtain insurance to cover such issues. Consequently, companies associated with a significant environmental incident may suffer significant financial losses. These losses are then translated into reductions in the share price and the associated dividends.

Banks that lend to companies secure the loans on the basis of the physical assets of the company and often on the land upon which any investment takes place. Should the company cease to be viable, the bank assumes ownership of those assets which are then sold to cover any outstanding debts. However, should the physical assets of the company be contaminated, then the value of the assets is significantly reduced. Indeed, the banks may inherit any environmental liabilities that the liquidated company generated. Commercial lenders are therefore reluctant to lend money to any company which may develop any environmental liabilities or to secure loans on the value of an asset which may be eroded through contamination. As a result, companies which cannot demonstrate a high level of environmental performance associated with low environmental risks will find it increasingly difficult and expensive to attract and retain investment and insurance for their operations.

Media and Pressure Groups

A combination of increased public awareness of environmental issues and freedom of access to information on the environmental performance of companies will serve to magnify media and pressure group interest in the environmental performance of industry. In order to manage media and pressure group attention, companies must be able to state that they have made efforts to reduce their environmental impact. However, while it may be tempting to allow the PR or marketing departments to lead the way in convincing all stake-holders of this commitment, any shallow or spurious claims will soon be uncovered. Claims which cannot be substantiated are likely to be seized upon and will be very detrimental to a company's public image. Companies which seek to communicate responsible environmental performance must base any claims that they make to this effect on hard facts which they are willing to communicate.

CLIMATE CHANGE FACTORS

Climate change is the result of a great many factors including the dynamic processes of the Earth itself, external forces including variations in sunlight intensity, and more recently by human activities. External factors that can shape climate are often called climate forcings and include such processes as variations in solar radiation, deviations in the Earth's orbit, and the level of greenhouse gas concentrations. There are a variety of climate change feedbacks that will either amplify or diminish the initial forcing. Most forms of internal variability in the climate system can be recognized as a form of hysteresis, where the current state of climate does not immediately reflect the inputs. Because the Earth's climate system is so large, it moves slowly and has time-lags in its reaction to inputs. For example, a year of dry conditions may do no more than to cause lakes to shrink slightly or plains to dry marginally. In the following year however, these conditions may result in less rainfall, possibly leading to a drier year the next. When a critical point is reached after "*x*" number of years, the entire system may be altered inexorably. In this case, resulting in no rainfall at all. It is this hysteresis that has been mooted to be the possible progenitor of rapid and irreversible climate change.

Plate Tectonics

On the longest time scales, plate tectonics will reposition continents, shape oceans, build and tear down mountains and generally serve to define the stage upon which climate exists. During the Carboniferous period, plate tectonics may have triggered the large-scale storage of Carbon and increased glaciation. More recently, plate motions have been implicated in the intensification of the present ice age when, approximately 3 million years ago, the North and South American plates collided to form the Isthmus of Panama and shut off direct mixing between the Atlantic and Pacific Oceans.

Solar Output

The sun is the source of a large percentage of the heat energy input to the climate system. Lesser amounts of energy is provided by the gravitational pull of the Moon (manifested as tidal power), and geothermal energy. The energy output of the sun, which is converted to heat at the Earth's surface, is an integral part of the Earth's climate. Early in Earth's history, according to one theory, the sun was too cold to support liquid water at the Earth's surface, leading to what is known as the Faint young sun paradox. Over the coming millennia, the sun will continue to brighten and produce a correspondingly higher energy output; as it continues through what is known as its "main sequence", and the Earth's atmosphere will be affected accordingly.

On more contemporary time scales, there are also a variety of forms of solar variation, including the 11-year solar cycle and longer-term modulations. However, the 11-year sunspot cycle does not appear to manifest itself clearly in the climatological data. Solar intensity variations are considered to have been influential in triggering the Little Ice Age, and for some of the warming observed from 1900 to 1950. The cyclical nature of the sun's energy output is not yet fully understood; it differs from the very slow change that is happening within the sun as it ages and evolves, with some studies pointing toward solar radiation increases from cyclical sunspot activity affecting global warming.

Solar variations are changes in the amount of solar radiation emitted by the Sun. There are periodic components to these variations, the principal one being the 11-year solar cycle (or sunspot cycle), as well as aperiodic fluctuations. Solar activity has been measured via satellites during recent decades and through 'proxy' variables in prior times. Climate scientists are interested in understanding what, if any, effect variations in solar activity have on the Earth. Effects on the earth caused by solar activity are called "solar forcing".

The variations in total solar irradiance remained at or below the threshold of detectability until the satellite era, although the small fraction in ultra-violet wavelengths varies by a few per cent. Total solar output is now measured to vary (over the last three 11-year sunspot cycles) by approximately 0.1% or about 1.3 W/m^2 peak-to-trough during the 11 year sunspot cycle.

The amount of solar radiation received at the outer surface of Earth's atmosphere averages 1,366 watts per square meter (W/m^2). There are no direct measurements of the longer-term variation and interpretations of proxy measures of variations differ.

On the low side North et al. report results suggesting ~ 0.1% variation over the last 2,000 years. Others suggest the change has been ~ 0.2% increase in solar irradiance just since the 17th century. The combination of solar variation and volcanic effects are likely to have contributed to climate change, for example during the Maunder Minimum. Apart from solar brightness variations, more subtle solar magnetic activity influences on climate from cosmic rays or the Sun's ultraviolet radiation cannot be excluded although confirmation is not at hand since physical models for such effects are still too poorly developed.

The longest recorded aspect of solar variations is changes in sunspots. The first record of sunspots dates to around 800 BC in China and the oldest surviving drawing of a sunspot dates to 1128. In 1610, astronomers began using the telescope to make observations of sunspots and their motions. Initial study was focused on their nature and behaviour. Although the physical aspects of sunspots were not identified until the 1900s, observations continued.

Study was hampered during the 1600s and 1700s due to the low number of sunspots during what is now recognized as an extended period of low solar activity, known as the Maunder Minimum. By the 1800s, there was a long enough record of sunspot numbers to infer periodic cycles in sunspot activity.

In 1845, Princeton University professors Joseph Henry and Stephen Alexander observed the Sun with a thermopile and determined that sunspots emitted less radiation than surrounding areas of the Sun. The emission of higher than average amounts of radiation later were observed from the solar faculae.

Around 1900, researchers began to explore connections between solar variations and weather on Earth. Of particular note is the work of Charles Greeley Abbot. Abbot was assigned by the Smithsonian Astrophysical Observatory (SAO) to detect changes in the radiation of the Sun. His team had to begin by inventing instruments to measure solar radiation. Later, when Abbot was head of the SAO, it established a solar station at Calama, Chile to complement its data from Mount Wilson Observatory.

He detected 27 harmonic periods within the 273-month Hale cycles, including 7, 13, and 39 month patterns. He looked for connections to weather by means such as matching opposing solar trends during a month to opposing temperature and precipitation trends in cities. With the advent of dendrochronology, scientists such as Waldo S. Glock attempted to connect variation in tree growth to periodic solar variations in the extant record and infer long-term secular variability in the solar constant from similar variations in millennial-scale chronologies.

Statistical studies that correlate weather and climate with solar activity have been popular for centuries, dating back at least to 1801, when William Herschel noted an apparent connection between wheat prices and sunspot records. They now often involve high-density global datasets compiled from surface networks and weather satellite observations and/or the forcing of climate models with synthetic or observed solar variability to investigate the detailed processes by which the effects of solar variations propagate through the Earth's climate system.

CHANGES IN WORLD TEMPERATURE

The climate has changed significantly in historic times. The little ice age that lasted from about 1550 to 1850 was a period when the global climate was cooler and winters were particularly cold. Since the 19th century there have been temperature data available from weather stations throughout the world. But it is difficult to compile an accurate picture of changes in world temperature over the last century due to various problems and errors. Overall there has been a worldwide warming of about 0.5 degree C since the end of the 19th century. During this time there have been two periods of rapid temperature increase, one between 1910 and 1930 and the other between 1970 and the present.

The 1990s was the warmest decade and 1998 the warmest year. On average, between 1950 and 1993, night-time daily minimum air temperature overland increased by about 0.2 degree C per decade. Warmest episodes of the El Nino Southern Oscillation (ENSO) phenomena have been more frequent, persistent and intense since the mid 1970s, compiled with the previous 100 years. In recent years, the large body of evidence that shows human activity is changing the global climate, raising temperatures and affecting ecosystems around the world. The Earth has warmed by approximately 0.75 degree C since pre-industrial times.

Global average temperatures have increased by 1.1 degree F over the last century - warming faster than any time in the last 1,000 years. As a result, the 1990s were the hottest decade in the past 1,000 years. There is overwhelming consensus that this is due to emissions of greenhouse gases, such as carbon dioxide (CO_2), from burning fossil fuels. Examination of ice cores shows that there is more CO_2 in the atmosphere than at any time in the past 600,000 years. Between 1960 and 2002, annual anthropogenic global emissions of CO_2 approximately tripled. They rose by about 33 per cent since 1987 alone.

Warming in this century is projected to be between 1.4 and 5.8 degree C.

After reviewing hundreds of studies that used data and climate models to examine past and future changes in climate extreme, a team of scientists, led by David R. Easterling of NOAA's National Climatic Data Centre reached a conclusion that our climate changes, extreme weather events such as droughts, floods, heat waves, heavy rainfall, tropical storms and hurricanes are expected to increase. This trend is likely to become more intense in the years to come both as the climate continues to change, and society continues to become more vulnerable to weather and climate extremes.

Scientists are increasingly concerned about the possibility of abrupt climate change, including reductions in ocean currents, such as the monsoon seasons, which would affect food security for billions of people. Today, most mainstream scientists and scientific bodies agree that heat-trapping gases like carbon dioxide - produced mainly from the burning of fossil fuels in cars, power plants factories and homes have caused temperatures to rise around the globe. Carbon dioxide, the major greenhouse gas, is currently found in the atmosphere in a concentration of 379 parts per million, and is almost certain to reach 540 ppm sometime in the next 40 to 50 years, according to the Intergovernmental Panel on Climate Change (IPCC). It could reach 800 to 1000 ppm before the end of the century. According to the report by the IPCC, a doubling of carbon dioxide in the atmosphere would produce a global temperature rise ranging from 3 to 8 degrees F. Humans are the dominant force behind the sharp global warming trend seen in the 20th century. The natural factors like volcanic eruptions and fluctuations in sunshine, which were powerful influences on temperatures in the past centuries, can account for only 25 per cent of the warming since 1900. The rest of the warming was caused by human activity, particularly rising levels of carbon dioxide and other heat-trapping gases.

The natural variability plays only a subsidiary role and the most parsimonious explanation for most of the warming is that it is due to the anthropogenic increase in greenhouse gases (GHGs). The impacts of climate change are already visible. Examples include: the shrinking Arctic ice cap; accelerating sea level rise; receding glaciers worldwide; thawing permafrost; earlier break-up of river and lake ice; increasing intensity and duration of tropical storms; lengthening of mid-to high-latitude growing seasons; and shifts in plant and animal ranges and behaviour.

Sea Level Rise

In the past 100 years, global sea level rose between 1 and 2 millimetres a year. Since 1992 the rate has increased to about 3 millimetres a year, primarily through thermal expansion of warming oceans and freshwater flowing into the oceans from melting ice. Melting ice is responsible for a significant portion of the observed sea level rise, with the Greenland and Antarctic ice sheets the largest contributors. The Greenland Ice Sheet is melting faster than new ice is being formed.

Scientists know that warm sea surface temperatures provide the energy source for tropical storms, but many other factors influence whether hurricanes form and which way they go. Scientists projected current climate conditions for 2080 using by nine different global climate models. Comparing the results, they found that global warming would increase hurricane intensity by 8 to 16 per cent, with rainfall increasing by 12 to 26 per cent within 60 miles of the storm centre. Meanwhile, sea level rise due to global warming will push shorelines inland by 400 feet or more in low-lying areas, making storm surges even more damaging.

Land Degradation

The severe droughts and forest fires of recent years in the regions of Mediterranean Europe, is threatening the livelihoods of 16.5 million Europeans. There are 300,000 square kilometres of territory currently affected. Degradation is mainly due to human activities such as over farming or land clearance, although drought also degrades the quality and productivity of soil and vegetation. The average temperature of major Asian cities could rise by 3 to 10 degrees Celsius by the end of the century while longer droughts and flooding threaten rural areas.

Antarctica

Antarctica is a continent surrounded by ocean. It is the coldest, windiest and driest continent on earth. Close to 99 per cent of the continent is covered by an ice-sheet with an average height above sea level of approximately 2,500 metres. Antarctic ice sheet has an average depth of around 2000 metres. The largest depth that has been measured is approximately 4,700 metres. The amount of ice in Antarctica constitutes about 91 per cent of the world's total. If all Antarctic ice were to melt, sea level would rise by more than 60 metres.

Antarctica has no native terrestrial vertebrates, but large populations of marine birds and seals which go there to breed. A large proportion of the world's seals are found in the ocean surrounding Antarctica. In the summer season both toothed whales and baleen whales are present in the Antarctic marine environment. Around 45 species of nesting birds are found south of the Antarctic Convergence, all of which are connected to the marine ecosystem. Penguins constitute 85 per cent of the biomass of Antarctic sea birds. In the Antarctic, three large sections of ice shelves in the Antarctic Peninsula have collapsed over the past 11 years.

Arctic Sea Ice

There are important geographical and political distinctions between the Arctic and Antarctic. The arctic is a partially-frozen ocean surrounded by a diversity of landscapes influenced by seasonal snow cover and permafrost, including ice, sparsely-vegetated barren lands, tundra, wetlands and forests. The Arctic Ice Cap consists of glaciers, ice sheets, icebergs and sea ice. Sea ice covers approximately 7.5 to 15 million square kilometres of the Arctic Ocean, with an average thickness of about three metres. During the summer, 10 to 15 per cent of the Arctic Ocean is not covered by ice. Greenland Ice Sheet is the largest Arctic glacial mass.

It constitutes 10 per cent of the world's total freshwater reserves. Melting of sea ice will not increase sea levels, but if all the Green ice were to melt, the sea level in the world's oceans would rise by seven metres. The Arctic is warming twice as fast as the global average.

NASA satellite images show a 20 per cent decline in summer arctic sea ice since 1979, while Antarctica glaciers are melting faster than expected accelerating sea level rise. Since 1980, between 20 and 30 per cent of sea ice in the European Arctic has been lost. Some climate models project that there may be an almost complete loss of summer sea-ice in the Arctic before the end of the century.

If this happens, polar bears are unlikely to survive as a species. The Arctic Climate Impact Assessment found that in Alaska, western Canada and eastern Russia average winter temperatures have increased as much as 40 to 70F in the past 50 years. Hunting has become more difficult and dangerous due to less reliable sea-ice conditions.

Highland Glaciers

In China, highland glaciers are shrinking each year by an amount equivalent to all the water in the Yellow River. The Chinese Academy of Science says that seven per cent of the country's glaciers are vanishing annually. By 2050, as many as 64 per cent of the country's glaciers will have disappeared. An estimated 300 million people live in China's west and depend on water from glaciers for their survival.

Coral Reefs

Coral reef is thought to host the most species-rich communities of the marine environment. They are comparable to tropical rainforest in that damage to their ecosystems may affect thousands of species adversely. Two thirds of all marine fish species are associated with tropical reefs where many human societies are depended on coral reefs for food, sport, protection of shorelines from storm damage, and tourism. All the nucleus of the coral reefs that have photosynthetic plant cells living symbiotically in their tissues.

This plant-animal partnership is responsible for an efficient recycling of nutrients. Geologically, reefs are made up of calcium carbonate produced by corals and other organisms, most notably coralline algae that has accumulated layer upon layer over thousands of generations. Reef communities are dependent on this reef structure for substrate and shelter and also on the reef's primary production, to which the corals and algae make substantial contribution.

In the tropical waters, no reef occurs where the waters are too deep, too muddy, too diluted by fresh water, or too hot. No coral reefs occur where temperatures exceed 30 degrees C for extended periods. It is also noted that many coral reefs are already near their upper thermal limit, at least for some months of the year. Thus, the small increases of 20-30 degrees C that are predicted for the surface waters of tropical oceans have profound implications for the structure, function, and distribution of reef ecosystems.

At least three major effects of temperature increase on coral reef may be identified. The first is physiological. Corals may expel their symbiotic algae in response to heat stress so that they appear bleached. Without those cells, corals cannot grow, and unless the algae become re-established, the corals die within a few months. Warming of tropical eastern Pacific waters contributed to wide-spread coral bleaching and death in 1982-83, and surface water temperatures above 30 degrees C are thought to have been responsible for the wider spread bleaching of corals in the Caribean Sea in 1987 (Roberts, 1987).

Second, mechanical damage to reefs could increase. If corals die and reef growth stops, the reef will become more vulnerable to erosion.

Further more, warming of tropical oceans may increase the frequency of hurricanes. Mechanical damage due to storms is a major source of coral mortality. Hurricanes can strip all living corals from long stretches of reef. Third, thermal expansion of ocean water, among other factors, causes elevation of sea level. The rate of sea level rise may affect the extent, structure and functioning of coral reef communities.

Elkhorn, staghorn and fused-staghon coral, which live in the Caribbean and off the southeast coast of Florida, have declined up to 98 per cent in the past 30 years, largely due to global warming. New research shows that carbon dioxide is a direct threat to corals because ocean water becomes more acidic (lower pH) as it absorb some of the CO2 emitted by power plants, automobiles and other sources. With corals already suffering from rising temperatures, the additional stress of falling pH could push them over the edge.

The oceans have absorbed approximately half of the CO2 produced in the past 2,000 years, producing carbonic acid and lowering the pH of surface seawater. This could affect the process of calcification by which animals such as corals and molluscs make their shells from calcium carbonate.

Agricultural Industry

Although crop yields may increase in some areas due to climate change, the negative effects are likely to dominate as warming increases. Africa is especially vulnerable, and studies warn that there may be a significant increase in hunger. Poor communities are most directly dependent for their livelihoods on a stable and hospitable climate. They often rely on rain-fed subsistence agriculture, and are deeply dependent on climatic phenomena, such as the Asian monsoons.

A comprehensive regional study of the impacts of global warming in California shows that higher temperatures and summer water shortages would seriously harm California's $30 billion agricultural industry. Alpine meadows in the Rocky Mountains to disappear, sugar maple trees to vanish in the Northeast, and greater risk from storm surges in the Southeast. Rising temperatures will also exacerbate water shortages in the west.

As glaciers melt in the world's great mountain ranges, water supplies to rivers will be affected. In Europe, eight out of nine glaciated regions show significant retreat. Between 1850 and 1980, glaciers in the European Alps lost approximately one-third of their area and one-half of their mass. This is also bad news for everyone who enjoys the fruit of their vines. The quality of California wine grapes would be degraded by excessive temperatures during ripening.

More Bad Air Days

Hotter temperatures caused by global warming will speed formation of the lung-damaging pollution commonly known as smog, significantly reducing the number healthy air days enjoyed by residents in more than a dozen cities in the USA. The residents of the inner city are particularly vulnerable to the affects of climate change and global warming.

The greatest threat is from heat waves. Exposure to excessive heat caused over 8,000 deaths in the United States between 1979 and 1999. Extreme heat waves caused more than 20,000 deaths in Europe and more than 1500 deaths in India during 2003. Higher temperatures also elevate the level of ozone smog in urban areas, which contributes to excess mortality and triggers more asthma attacks.

The scientific findings released over the past few years show the humankind's impact on earth's climate. Given this growing body of evidence, we must act now to reduce pollution from vehicles and power plants. By deploying already available technologies it is possible to prevent a doubling of carbon dioxide in the atmosphere during the next 50 years and avoid the most dangerous threats from global warming. The British Government has committed to reducing its emissions of heat-trapping gases by 60 per cent from 1990 levels by mid-century and urging other industrialized countries to adopt the same goal.

Any delay would be dangerous, because an additional warming of merely one degree Celsius could be enough to trigger the eventual disintegration of ice sheets in Greenland and parts of Antarctica. The international community needs to work together immediately, not only to stabilise the level of heat-trapping greenhouse gases, but also to develop alternative technologies in order to move away from our dependence on fossil fuels. Delaying action will only make it "more disruptive and more expensive" to deal with global warming. Substantial reductions in emissions can come from improvements in fuel efficiency of cars and trucks, policies that require energy efficiency and the use of renewable energy, and long-term investments in new technologies like hydrogen fuels and bio fuels.

Carbon dioxide is the main greenhouse gas (GHG), and slowdown of its emissions must have priority. It will be a growing issue in international relations for decades. 'Kyoto' gives too little or no weight to gases such as methane, the trace gas HFC-134a, ozone and the precursor gases that form ozone. The slowdown in the growth rate of the GHGs contribution to global warming from the peak in the 1980s is due mainly to the phase out of CFCs as dictated by the Montreal Protocol. This success could be diminished by increases of other trace gases not controlled by the Montreal Protocol. Therefore, it is well worth extending the Montreal Protocol machinery to phase out many of these trace gases.

More than half of all humanity has probably always lived within coastal areas of the world, and that proportion is increasing rapidly. Human activities have already caused extensive alteration of natural coastal ecosystems, which can ill afford further damage. Therefore, natural and international efforts to protect coastal areas should be given the highest priority, and long-term plans should include provisions for dealing with climate change. Climate change cause profound changes in the ecology of marine systems, but also climate change will add stress to systems that are already experiencing increasing and often severe disruption from other human activities, including pollution, habitat destruction, and over harvesting of the seas.

All of the changes that we have mentioned have consequences not only for species and ecological diversity but also for the human economic future.

Change in the abundances or geographical availability of fishery resources, for example, will have great human impacts. The consequences will be ever larger if warming affects basic ecological processes such as primary productivity, reef building, and the lifestyles communities."

Measurements by four major temperature tracking outlets reported that world temperatures dropped by about 0.65° C to 0.75° C during 2007, the fastest temperature changes ever recorded (either up or down). The cooling approached the total of all warming that occurred over the past 100 years, which is commonly estimated at about 1° C. Antarctic sea ice expanded by about 1 million square kilometers – more than the 28-year average since altimeter satellite monitoring began. But have these collective announcements ended the global warming debates? No, stay tuned for further developments.

Cyclical, abrupt, and dramatic global and regional temperature fluctuations have occurred in observable patterns over millions of years, long before humans invented agriculture, capitalism, smokestacks, and carbon trading schemes. To appreciate just how lucky we are to live in the present, consider climate cycles from a historical perspective. Over the past 400,000 years, much of the Northern Hemisphere has been covered by ice up to three miles thick, at regular intervals lasting about 100,000 years each. Very brief interglacial cycles lasting about 12,000 to 18,000 years, like our current one, have offered reprieves from the bitter cold. From this perspective, there can be no doubt that current temperatures are abnormally warm. The average temperature of our planet has been gradually increasing on a fairly constant basis over the past 18,000 years or so since it began thawing out of the last ice age. About 12,000 to 15,000 years ago, the Earth had warmed enough to halt the advance of the glaciers and cause sea levels to rise. About 8,000 years ago, a land bridge across the Bering Strait submerged, cutting off migrations of people and animals to North America.

CLIMATE CHANGE—BEYOND WITHERING WEATHER

Climate change is about much more than how warm or cool our temperatures are. Whereas "global warming" refers to increasing global temperatures, "climate change" refers to regional conditions. Climate is defined by a number of factors, including:

- Average regional temperature as well as day/night temperature patterns and seasonal temperature patterns.
- Humidity.
- Precipitation (average amounts and seasonal patterns).
- Average amount of sunshine and level of cloudiness.
- Air pressure and winds.

- Storm events (type, average number per year, and seasonal patterns).

To a great extent, this is what we think of as "weather." Indeed, weather patterns are predicted to change in response to global warming:

- Some areas will become drier, some will become wetter;
- Many areas will experience an increase in severe weather events like killer heat waves, hurricanes, flood-level rains, and hail storms.

It's tempting to think that all of these changes to the world's climate regions will average out over time and geography and things will be fine. In fact, colder climates like Canada may even see improved agricultural yields as their seasonal temperatures rise. But overall, humanity has made a huge investment in "things as they are now, where they are now." Gone are the days of millennia ago when an unfavorable change in climate might cause a village to pack up their relatively few belongings and move to a better area. We have massive societal and industrial infrastructure in place, and it cannot be easily moved.

Climate-change effects will generally not be geographically escapable in the timeframe over which they happen, at least not for the majority of humans and species. Beyond mere "weather," we can also think of a region's climate as a place in which things live—a la, "a hospitable climate." Therein lies the real danger of global warming—climate change will affect the success or failure of how plants and animals live in a given geographic area, including food crops.

We think of the Central Valley of California as a lush, agriculturally productive landscape, but central California's climate is actually quite dry. Without intensive use of irrigation, the land would not produce the volume or variety of food it does now. So, what if increasing temperatures cause less snow pack to accumulate in the mountains each year, leading to lower river flows and less water available for irrigation in California's agricultural areas? What if changes in rainfall patterns make central California's climate even drier? How much would crop output fall?

This is just one example of how global warming could lead to a regional climate change that would cause a big difference in local economics and the national food supply. Even though the main threat right now is warming planetary temperatures, climate change can also mean global cooling.

The last Ice Age was part of a globally cool period and it featured some rather severe "climate change" characteristics. It's worth remembering that global warming is based on an increasing average global temperature.

Some parts of the planet (such as the Arctic) are getting warmer much faster than other areas. It's even possible that some regions could actually experience regional cooling at the same time the planet as a whole is experiencing global warming. Here's how. The "thermohaline circulation" in the world's oceans is part of the planet's temperature regulation system. It can warm or cool regional climates to make their average temperatures different that they would be normally based on their latitude.

The most notable example of this is how the Gulf Stream brings warm water up from the tropics to make Europe much warmer than it would be naturally. This part of the thermohaline circulation is dependent on regular additions of fresh water from melting Arctic ice (which is replaced every year through additional snowfall). If the flow of fresh Arctic water decreases enough, it could slow or even stop the thermohaline circulation, leading to cooler temperatures in Europe—even at the same time other areas are experiencing severe temperature increases.

There are important differences between global warming and climate change, but the two are closely intertwined in determining the climate futures for the regions where we live. Predictions of regional impacts are beginning to emerge from climate models. There are regions that will get some benefits, but most of the predicted effects around the world are bad—bad enough that we need to quickly start fixing our greenhouse gas emissions.

There are solutions, but most leaders today are offering only low-impact solutions that will not truly solve the problem. We need to be talking about how to cut greenhouse gas emissions by 50%-80% over the next few decades, not dithering over minimalist efforts like how to get back to 1990 levels by the year 2020.

Some leaders just offer excuses about why no action is possible, citing "the economy" as a reason to continue ignoring the problem. But a report from the British government about the economic damage that will be caused by inaction on climate change makes it clear that continued delay is unwise, even from an economic standpoint. Many of the actions necessary to solve the global warming problem must be attacked at the national and international levels. But in the US, states and cities are thus far in the lead on implementing solutions. In the end, total success will have to be built on our actions as individuals (regardless of country), from energy conservation to vehicle choices to what we demand of our leaders. You too can be part of these global warming solutions!

Global warming fears aside, all students of climate science know that the Earth is presently in an Ice Age and has been for approximately the past 2 to 2.5 million years. This Ice Age has been characterized by successive advances and retreats of a glacial ice sheet, originating in Greenland and extending across the northern portions of the North American and Eurasian continents. Just 12,000 years ago, the undisputed geological evidence shows that New

York, Chicago, and all of North America up to the Arctic regions were under a sheet of ice, estimated to have been from 1 to 2 miles thick. Mountain glaciers also extended downward from the Rocky Mountains and the Appalachians in regions further to the south than the main glacial mass.

A similar situation prevailed over most of Germany, northern France, the British Isles, Scandinavia, Poland and other parts of Eastern Europe, and Russia. Such had been the state of things on Earth for probably at least 100,000 years. Before that, a short period known as an interglacial had allowed for a warm climate somewhat like the present, and before it another extended period of glacial advance. The thaw which produced our present geography—the Great Lakes, the southward flowing Ohio River, and much else we take for granted—was not completed until about 8,000 to 9,000 years ago, according to the best estimates of geologists and climatologists. Not only were there changes in the internal geography, but the continental boundaries were also greatly changed during the glacial period.

Calculations of the volume of water that must have been contained frozen in the continental glaciers, indicate that the global sea level, was lower by as much as 300 to 400 feet at times of glacial advance. A glaciation does not mean sea level rise, but a sharp fall in sea level exposing the continental shelf for miles out to sea. Much of the coastal-dwelling civilization of the past 100,000 or more years, thus lies buried offshore beneath hundreds of feet of ocean. Can this happen again? The most plausible theory of the causes of the ice ages, the theory of astronomical determination, suggests that the time is ripe for it to happen sometime soon. A Jan. 11 article in the online edition of the Russian daily Pravda was titled "Earth on the Brink of an Ice Age."

Many recent signs, including the recent deadly cold wave in Europe, and an extended period of reduced sunspot activity known as a solar minimum, which can contribute to the onset of glaciations, indicate that the Earth may be headed into a period of serious cooling. Perhaps it will be the beginning of a period of several hundred years duration, known as a Little Ice Age, perhaps the onset of a full-scale glacial advance to last for another 100,000 years. Here's what the Russians are talking about. The Orbital Theory of Climate The best available theory for explaining the advance and retreat of the northern ice sheets is that they are driven by changes in the orbital relationship of the Earth to the Sun, which affect the amount of solar radiation reaching the high northern regions.

There are three major cycles of change in the Sun-Earth relationship, all of them related to the discovery of Johannes Kepler that the orbit of the Earth is not a circle, but an ellipse with a difference between nearest and farthest approach to the Sun of about 3 to 4 million miles. The first of these cycles, known as the precession of the equinox, was known to the ancient Vedic astronomers. It is a cycle of approximately 26,000 years, produced by the wobble of the Earth's axis. Corrected for another phenomenon known as the advance of the perihelion, it means that about every 21,000 years, the direction of the tilt of the Earth's axis is such that Summer in the northern hemisphere is occurring when Earth finds itself at the point in its orbit most distant from the Sun.

The two other important cycles are: First, the change in the angle of inclination of the Earth's axis, known as obliquity, which varies on a 40,000 year cycle from about 22 to 24.5 degrees. Second, the variation in the dimensions of the elliptical orbit itself, which stretches like a rubber band from more like circular to more elliptical. This cycle, known as ellipticity, is more complicated to estimate, but peaks about every 100,000 years. The suggestion of an astronomical climate driver was first advanced in 1830 by the English astronomer John Herschel, son of the German musician Wilhelm Herschel who emigrated to England in the 18th century and founded modern observational astronomy.

Studies of the advance and retreat of Alpine glaciers next prompted the Swiss-born American scientist Louis Agassiz to suggest that such changes might occur on a broader scale forming continental ice sheets which could account for many hitherto unexplained geological phenomena. Attempts at developing an astronomical causation for such large-scale glacial activity were tried by French mathematician Joseph Adhemar (1842) and later in the 19th Century by the Scottish autodidact James Croll. The modern version of the theory originated with the astronomer-meteorologist Vladimir Köppen. Born in St. Petersburg and educated at Heidelberg, Köppen was employed at the German Marine Observatory at Hamburg when he first recognized that it was not an extremely cold Winter, but rather a cool Summer that was required for glacial advance.

If, during the short Summers in the high northern latitudes, the amount of solar radiation was insufficient to melt back the snow and ice that had formed over the Winter period, there would be glacial advance. Allow this to continue for year after year, and a sheet of ice might develop its own momentum, utilizing its high reflectivity for sunlight as a protective shield for maintaining cool surrounding airs. With his son-in-law, Alfred Wegener, better known as the prescient author of the theory of continental drift, the two began to conceptualize the changes in the amount of solar radiation (insolation) which would occur as the three cycles of orbital variation worked together to reinforce or cancel one another.

In order to see how a glaciation might begin, conceive of the Earth-Sun relationship as such that while the ellipticity of the orbit is at a maximum (meaning that aphelion is occurring at the greatest possible distance from the Sun), the Earth's axis is so oriented in the precession cycle that northern hemisphere. Summer is also occurring at aphelion. This is the present orbital position of the Earth with respect to the Sun. The resultant reduction in insolation will then make it possible that the snow and ice accumulation which occurred the previous. Winter does not melt back fully. Add to that the effect of a lessened axial tilt (obliquity), which reduces the amount of Summer insolation, further contributing to the growth of the ice sheet.

In 1920, Köppen enlisted the support of Serbian mathematician Milutin Milankovitch who worked out the astronomical theory of climate with mathematical precision, predicting when the overlying cyclical waves of precession, obliquity and ellipticity would tend to reinforce or to cancel one another. His results suggested that the 40,000-year cycle would be the dominant one. Can We Date the Ice Ages? Following improvements in the ability to measure isotope ratios which came about as a spin-off of the wartime Manhattan project, physical chemist Harold Urey began to examine the possibility that the ratio of the two principal isotopes of oxygen found in the atmosphere might provide a clue as to past temperatures.

It was based on the idea that the ratio of the heavier isotope (oxygen-18) to the more prevalent isotope (oxygen-16) found at the sea surface would change depending on the temperature of the ocean water near the surface. Urey thought that a careful study of the oxygen isotope ratio in the shells of sea creatures, which build their calcium carbonate shells from oxygen available in the seawater, might serve to indicate the temperature of the water in which they formed. During warmer periods, it was thought, evaporation from the ocean surface would tend to enrich the sea surface water with the heavier isotope of oxygen.

Perhaps, Urey reasoned, the isotope ratios found in the layers of discarded shells of sea organism which form the ocean bottom could thus serve as a record for the past temperatures of the ocean. The theory is fraught with many ifs, but it was pursued with persistence, starting in the 1950s, by Italian-educated micropaleontologist Cesare Emiliani, a one-time collaboator of Urey at the Argonne Laboratory then associated with University of Chicago. Emiliani identified certain species of small shell-forming sea organisms known as foraminifera, which he thought suitable for oxygen-isotope analysis to determine past climates.

The conclusions he drew as to the dating of the ice ages were constantly challenged by leading oceanographers, who found them in contradiction with their studies of ocean bottom cores. The method was also attacked on the grounds that it wasn't clear that the creatures formed their shells, known as tests, near enough to the surface to reflect changes in isotope ratios. About 1968, a somewhat new interpretation of the oxygen isotope record was proposed by a young oceanographer and climatologist, Nicholas Shackleton, a Cambridge graduate and great nephew of the famous British Antarctic explorer of the same name.

Shackleton proposed that the oxygen-isotope ratio could serve as a proxy, not for temperature but for sea level—the idea being that during periods of glacial advance, when a large volume of ocean water had been taken up into the continental ice sheets, the oxygen-18 ratios of the remaining water would consequently be higher. These might be detected in the foraminifera layers found in the ocean bottom cores. Again there are many ifs, but Shackleton examined isotopic ratios of snows in Alpine and Arctic regions as well as many other factors to bolster his hypothesis. In the 1970s a National Science Foundation-funded programme of oceanographic studies, known as CLIMAP, collected a large number of sediment cores from different parts of the world ocean.

The programme, known as the Decade of the Oceans, was run in conjunction with some flawed statistical approaches to modeling of global atmospheric circulation that had originated in efforts of John von Neumann to use computer modeling for studies of weather modification. However, analysis of the oxygen isotopic ratios of foraminifera found in the undersea cores suggested to a team working at the Lamont-Doherty Geological Laboratory that there was a definite signal of 100,000 year cyclicity. Dr. John Imbrie, who ran the computer programs analyzing the data, was the first to hypothesize that the periodicities were caused by the Milankovitch orbital cycles.

A landmark paper by Hays, Imbrie and Shackleton, published in the December 1976 issue of Science magazine ("Variations in the Earth's Orbit: Pacemaker of the Ice Ages"), argued that the advance and retreat of the ice sheets was triggered by the changes in the Earth's orbital parameters. Other factors might also be present to reinforce these relatively small changes in solar radiation, but these were the pacemaker. By the theory of the orbital cycles, the evidence from the undersea cores explained that a major glaciation would be set off about every 100,000 years, followed by a short period known as an interglacial, a melt back lasting about 10,000 to 12,000 years.

By the calculations of astronomers, the present interglacial, which has lasted about 11,000 years, is due to end any time. Indeed we have been in a period of long-term cooling for more than 6,000 years. The maximum summer temperature experienced in Europe over the last 10,000 years occurred about 6000 B.C. Over North America, where the process of glacial retreat lagged somewhat, the maximum was reached by about 4000 B.C. These estimates based on a vast array of evidence from geology, botany, and many other fields are consistent with the orbital theory of climate, for the northern hemisphere Summer would have been occurring at a point in Earth's orbit much nearer to the Sun than presently.

Where Global Warming Came From The onset of a glaciation leading to the formation of a new ice sheet can be very sudden, according to the paleoclimatic evidence. Shifts from warm to cold phases may also come on surprisingly fast, as occurred during the recent period known as the Little Ice Age that persisted from about 1300 to 1850, and may be occurring now. If the theory of orbital determination is correct, we should thus take the threat of a new ice age very seriously. And for a short time in the 1970s, we did. However, other forces were at work. The intent of the shapers of global policy grouped around the banner of Anglo-Dutch financial oligarchy, was to use the thaw in U.S.-Soviet relations that had been signaled by the limited nuclear Test Ban Treaty of 1963 to accomplish the phased destruction of the scientific research capabilities of the leading powers, especially the United States.

The manipulation of popular opinion against science, first by the Bertrand Russell-led nuclear weapons test scare, then by a succession of environmental hoaxes originating with the widespread circulation of Rachel Carson's 1962 Silent Spring, was to be the leading weapon in this assault on the idea of scientific progress itself. In 1975, a year before the publication of the paper promoting the orbital theory of climate, a conference organized by then president of the American Association for the Advancement of Science, Dame Margaret Mead, had sealed the fate of the astronomical theory of climate.

Mead and conference co-organizer William Kellogg (a climate scientist from RAND and later NCAR, the National Centre for Atmospheric Research) had determined that the now well-known theory of carbon dioxide-induced global warming was to become the official interpretation of climate phenomena. Scares about a new ice age were all very well, but to really channel popular opinion against scientific development, it was much more effective to blame modern industrial activity—the carbon dioxide produced from burning of fossil fuels—for the danger. Titled "The Atmosphere: Endangered and Endangering," the conference theme was that scientists must not worry so much about accuracy and complex ideas, but streamline, simplify, and if necessary distort their results in order to more effectively mobilize policy makers and public opinion against the alleged dangers of greenhouse gases.

It was a naked attempt to misuse science to drive the real agenda of reducing population by stopping the spread of scientific and industrial progress. Among the leading participants were three top students of Malthusian

Paul Ehrlich. One of them Stanford University climatologist and global warming scaremonger Stephen Schneider, later became notorious by carrying the spirit of the conference into a 1989 interview with Discover magazine: "To capture the public imagination, we have to offer up some scary scenarios, make simplified dramatic statements and little mention of any doubts one might have. Each of us has to decide the right balance between being effective, and being honest."

This was the origin of Al Gore's campaign. Prior to that, the theory of anthropogenic global warming was no more than an also-ran. Although the notion that the carbon dioxide output of human industry might affect global climate had been proposed in the 1890s, repeated attempts to measure its effects had shown no significant influence. To generate a global warming scare required a mobilization of resources and funding to universities and institutions on an unprecedented scale. By the early 1990s expenditures to aid they were reaching into the billions of dollars a year; from 2001 to 2007 annual government funding for to the global warming scare industry had reached $5 billion. A brief episode of warming over some parts of the globe from about the mid-1970s to 1998 helped to feed the scare among a gullible and increasingly science-deprived population.

That is now over. In the decade since 1998, the averaged global temperature has fallen about 0.6 degrees C, canceling the entire increase in average temperature that had been claimed for the prior century, based on microwave sensor satellite data from the Marshall Spaceflight Centre analyzed by Drs. John Christie and Roy Spencer at the University of Alabama in Huntsville. Recently, the index of sunspot activity, a measure which correlates with the output of heat radiation from the Sun, has fallen to lows not seen since 1913.

Apart from changes in the orbital configuration, the output of the Sun itself affects climate, both directly as heat and through indirect means. For example, the solar wind, the output of charged particles form the Sun, affects the influx of cosmic rays. Cosmic rays, it has recently been demonstrated, may play a crucial part in generating the seed crystals around which clouds form. More cosmic radiation entering the Earth's atmosphere, because of a weaker Sun, may mean more cloud cover and more global cooling.

The Sun is known to go through 11-year cycles of increase and decrease in its output. But the recent cycle has been so weak that some specialists fear we are entering another Dalton Minimum, the period from 1790 to 1830—in the midst of the Little Ice Age—when sunspot activity was at a similar low.

Astronomer Khabibullo Abdusamatov of Russia's Pulkovo Observatory predicted in 2005 that solar activity was about to decline, and a new cooling would ensue. Last year, Russian Academician Dr. Oleg Sorokhtin advised the world to "stock up on fur coats." Sorokhtin predicted the occurrence of a solar minimum by the year 2040, and a prolonged period of glaciation following.

The intent of the global warming scare is to reduce world population. That is the stated intent of its initiators and of such important promoters of this and related anti-science scares as the World Wildlife Fund, founded by Britain's Malthusian Prince Philip and former Nazi SS officer Prince Bernhard of the Netherlands.

Isn't it time to stop being a sucker for the fear- mongering of people with another agenda—a very evil agenda whose end result will be the reduction of world population from its current level of 5 to 6 billion to less than 2 billion souls, with the concomitant dissolution into perpetual war, famine and human misery that such a plan must entail? A return to the American System concept of science-driven progress will assure us that we have the best available means to meet any future challenge to human survival, whether from climate change, new disease organisms, or some threat as yet unforeseen. An Ice Age brought on by global warming was the scenario depicted in the movie THE DAY AFTER TOMORROW.

While the science on which the movie based has been called into question, there may be some merit in the theory that global warming could cause an Ice Age. Why is Europe's climate comparatively milder than other places at the same latitude? Alaska and Greenland, both the same distance to the North Pole as Europe, are covered with ice and permafrost while most of Europe is not. The ocean currents called the Gulf Stream bring warm waters up to Europe from the Caribbean. This water brings warmth to the countries in its path. Cooler water from Europe feeds back into the loop and causes the water to flow back to the Caribbean in a continuous cycle.

The Gulf Stream has been significantly weakened in every major cooling event, including the last great Ice Age. In the past this weakening has been brought on by natural events. In current times, global warming brought on by human activities could be the cause of slowing or even stopping the Gulf Stream. If this were to happen, the cold waters would stay in the area of Europe and the Northeastern US and could mean an Ice Age for those regions. If an Ice Age occurs, it will likely be due to the melting of polar ice. This will dump large quantities of cold, fresh water into the ocean.

It would disrupt the Gulf Stream and cause the cooling of many areas that now have milder climates. The return flow of cold water from Greenland, which goes back to the Caribbean, has already showed a weakening over the last 50 years. There has been a twenty per cent decline in the amount of current flowing in this direction. It stands to reason that the warm waters returning from the Caribbean have also decreased in volume.

The change would not be gradual. This is a phenomenon that takes place rather quickly. Perhaps it does not happen as fast as depicted in THE DAY AFTER TOMORROW. However, it could happen within a few short years. A slowing or stoppage of the Gulf Stream would affect the entire earth. Observations have been made of current data and historical information gleaned by studying the ocean and the lands around it. With all the information at hand, it appears that it is indeed possible that global warming could bring about a modern Ice Age.

"The phrase 'Global Warming' has become familiar recently as environmental issues have hit the headlines. Within the past decade, there has been a considerable rise in public awareness and interest in climate change and in the possible impact of human activities on global climate. Such interest is particularly acute amongst environmental scientists, but is also prevalent amongst politicians, economists and some industrialists and has resulted in continuing attention in the media. The focus of attention has been upon the present and continuing effects upon climate of both domestic and industrial fuel use and of other processes both agro-environmental and industrial that lead to the release into the atmosphere of so-called 'greenhouse' gases.

A recorded rise in temperature this century is mainly attributable to human activities and this rise will both accelerate and continue until well after greenhouse gas emissions are stabilised. Records show global temperatures increasing: consolidated data from selected terrestrial stations and marine sea-surface temperature (SST) records round the world reveal that the seven warmest years this century have occurred since 1980. The recorded rise in temperatures has been linked to an estimated 30 per cent rise of carbon dioxide concentrations in the lower atmosphere over the last 200 years. As the carbon dioxide concentrations in the atmosphere continue to rise, there might be major temperature rises and perhaps even catastrophic climate changes in the next century.

PHYSICAL EVIDENCE FOR CLIMATIC CHANGE

Evidence for climatic change is taken from a variety of sources that can be used to reconstruct past climates. Most of the evidence is indirect—climatic changes are inferred from changes in indicators that reflect climate, such as vegetation, ice cores, dendrochronology, sea level change, and glacial geology.

Glacial Geology

Glaciers are recognized as being among the most sensitive indicators of climate change, advancing during climate cooling (for example, during the period known as the Little Ice Age) and retreating during climate warming on moderate time scales. Glaciers grow and shrink, both contributing to natural variability and amplifying externally forced changes. A world glacier inventory has been compiled since the 1970s. Initially based mainly on aerial photographs and maps, this compilation has resulted in a detailed inventory of more than 100,000 glaciers covering a total area of approximately 240,000 km^2 and, in preliminary estimates, for the recording of the remaining ice cover estimated to be around 445,000 km^2. The World Glacier Monitoring Service collects data annually on glacier retreat and glacier mass balance From this data, glaciers worldwide have been shown to be shrinking significantly, with strong glacier retreats in the 1940s, stable or growing conditions during the 1920s and 1970s, and again increasing rates of ice loss from the mid 1980s to present. Mass balance data indicate 17 consecutive years of negative glacier mass balance. The most significant climate processes of the last several million years are the glacial and interglacial cycles of the present age. The present interglaciation (often termed the Holocene) has lasted about 10,000 years. Shaped by orbital variations, earth-based responses such as the rise and fall of continental ice sheets and significant sea-level changes helped create the climate. Other changes, including Heinrich events, Dansgaard–Oeschger events and the Younger Dryas, however, illustrate how glacial variations may also influence climate without the forcing effect of orbital changes.

Advancing glaciers leave behind moraines that contain a wealth of material - including organic matter that may be accurately dated - recording the periods in which a glacier advanced and retreated. Similarly, by tephrochronological techniques, the lack of glacier cover can be identified by the presence of soil or volcanic tephra horizons whose date of deposit may also be precisely ascertained. Glaciers are considered one of the most sensitive climate indicators by the IPCC, and their recent observed variations are considered a prominent indicator of impending climate change.

Vegetation

A change in the type, distribution and coverage of vegetation may occur given a change in the climate; this much is obvious. However, to what extent particular plant life changes, dies or thrives, depends largely on the model of prediction used. In any given scenario, a mild change in climate may result in increased precipitation and warmth, resulting in improved plant growth and the subsequent sequestration of airborne CO_2. Larger, faster or more radical changes, however, may well result in vegetation stress, rapid plant loss and desertification in certain circumstances.

Ice Cores

Analysis of ice in a core drilled from a permafrost area, such as the Antarctic, can be used to show a link between temperature and global sea level variations. The air trapped in bubbles in the ice can also reveal the CO_2 variations of the atmosphere from the distant past, well before modern environmental influences. The study of these ice cores has been a significant indicator of the changes in CO_2 over many millennia, and continues to provide valuable information about the differences between ancient and modern atmospheric conditions.

Dendrochronology

Dendochronology is the analysis of tree ring growth patterns to determine the age of a tree. From a climate change viewpoint, however, Dendochronology can also indicate the climatic conditions for a given number of years. Wide and thick rings indicate a fertile, well-watered growing period, whilst thin, narrow rings indicate a time of lower rainfall and less-than-ideal growing conditions.

Pollen Analysis

Palynology is the science that studies contemporary and fossil palynomorphs, including pollen. Palynology is used to infer the geographical distribution of plant species, which vary under different climate conditions. Different groups of plants have pollen with distinctive shapes and surface textures, and since the outer surface of pollen is composed of a very resilient material, they resist decay. Changes in the type of pollen found in different sedimentation levels in lakes, bogs or river deltas indicate changes in plant communities; which are dependent on climate conditions.

Insects

Remains of beetles are common in freshwater and land sediments. Different species of beetles tend to be found under different climatic conditions. Given the extensive lineage of beetles whose genetic makeup has not altered significantly over the millennia, knowledge of the present climatic range of the different species, and the age of the sediments in which remains are found, past climatic conditions may be inferred.

Sea Level Change

Climate models for the substantiation of theories regarding global warming rely heavily on the measurement of long-term changes in global average sea level. Global sea level change for much of the last century has generally been estimated using tide gauge measurements collated over long periods of time to give a long-term average. More recently, altimeter measurements — in combination with accurately determined satellite orbits — have provided an improved measurement of global sea level change.

Current sea level rise has occurred at a mean rate of 1.8 mm per year for the past century, and more recently at rates estimated near 2.8 ± 0.4 to 3.1 ± 0.7 mm per year (1993-2003). Current sea level rise is due partly to human-induced global warming, which will increase sea level over the coming century and longer periods. Increasing temperatures result in sea level rise by the thermal expansion of water and through the addition of water to the oceans from the melting of continental ice sheets.

Thermal expansion, which is well-quantified, is currently the primary contributor to sea level rise and is expected to be the primary contributor over the course of the next century. Glacial contributions to sea-level rise are less important, and are more difficult to predict and quantify. Values for predicted sea level rise over the course of the next century typically range from 90 to 880 mm, with a central value of 480 mm.

Based on an analog to the deglaciation of North America at 9,000 years before present, some scientists predict sea level rise of 1.3 meters in the next century. However, models of glacial flow in the smaller present-day ice sheets show that a probable maximum value for sea level rise in the next century is 80 centimeters, based on limitations on how quickly ice can flow below the equilibrium line altitude and to the sea.

Local mean sea level (LMSL) is defined as the height of the sea with respect to a land benchmark, averaged over a period of time (such as a month or a year) long enough that fluctuations caused by waves and tides are smoothed out. One must adjust perceived changes in LMSL to account for vertical movements of the land, which can be of the same order (mm/yr) as sea level changes. Some land movements occur because of isostatic adjustment of the mantle to the melting of ice sheets at the end of the last ice age.

The weight of the ice sheet depresses the underlying land, and when the ice melts away the land slowly rebounds. Atmospheric pressure, ocean currents and local ocean temperature changes also can affect LMSL. "Eustatic" change (as opposed to local change) results in an alteration to the global sea levels, such as changes in the volume of water in the world oceans or changes in the volume of an ocean basin.

Various factors affect the volume or mass of the ocean, leading to long-term changes in eustatic sea level. The two primary influences are temperature (because the volume of water depends on temperature), and the mass of water locked up on land and sea as fresh water in rivers, lakes, glaciers, polar ice caps, and sea ice. Over much longer geological timescales, changes in the shape of the oceanic basins and in land/sea distribution will affect sea level.

Observational and modelling studies of mass loss from glaciers and ice caps indicate a contribution to sea-level rise of 0.2 to 0.4 mm/yr averaged over the 20th century.

Glaciers and Ice Caps

Each year about 8 mm (0.3 inch) of water from the entire surface of the oceans falls into the Antarctica and Greenland ice sheets as snowfall. If no ice returned to the oceans, sea level would drop 8 mm every year. To a first approximation, the same amount of water appeared to return to the ocean in icebergs and from ice melting at the edges. Scientists previously had estimated which is greater, ice going in or coming out, called the mass balance, important because it causes changes in global sea level.

High-precision gravimetry from satellites in low-noise flight has since determined Greenland is losing millions of tons per year, in accordance with loss estimates from ground measurement. Ice shelves float on the surface of the sea and, if they melt, to first order they do not change sea level. Likewise, the melting of the northern polar ice cap which is composed of floating pack ice would not significantly contribute to rising sea levels. Because they are fresh, however, their melting would cause a very small increase in sea levels, so small that it is generally neglected. It can however be argued that if ice shelves melt it is a precursor to the melting of ice sheets on Greenland and Antarctica.

- Scientists previously lacked knowledge of changes in terrestrial storage of water. Surveying of water retention by soil absorption and by reservoirs outright ("impoundment") at just under the volume of Lake Superior agreed with a dam-building peak in the 1930s-1970s timespan. Such impoundment masked tens of millimeters of sea level rise in that span.
- If small glaciers and polar ice caps on the margins of Greenland and the Antarctic Peninsula melt, the projected rise in sea level will be around 0.5 m. Melting of the Greenland ice sheet would produce 7.2 m of sea-level rise, and melting of the Antarctic ice sheet would produce 61.1 m of sea level rise. The collapse of the grounded interior reservoir of the West Antarctic Ice Sheet would raise sea level by 5-6 m.
- The snowline altitude is the altitude of the lowest elevation interval in which minimum annual snow cover exceeds 50%. This ranges from about 5,500 metres above sea-level at the equator down to sea level at about 70° N&S latitude, depending on regional temperature amelioration effects. Permafrost then appears at sea level and extends deeper below sea level polewards.
- As most of the Greenland and Antarctic ice sheets lie above the snowline and/or base of the permafrost zone, they cannot melt in a timeframe much less than several millennia; therefore it is likely that they will not, through melting, contribute significantly to sea level rise in the

coming century. They can, however, do so through acceleration in flow and enhanced iceberg calving.

- Climate changes during the 20th century are estimated from modelling studies to have led to contributions of between –0.2 and 0.0 mm/yr from Antarctica (the results of increasing precipitation) and 0.0 to 0.1 mm/yr from Greenland (from changes in both precipitation and runoff).
- Estimates suggest that Greenland and Antarctica have contributed 0.0 to 0.5 mm/yr over the 20th century as a result of long-term adjustment to the end of the last ice age.

The current rise in sea level observed from tide gauges, of about 1.8 mm/yr, is within the estimate range from the combination of factors above but active research continues in this field. The terrestrial storage term, thought to be highly uncertain, is no longer positive, and shown to be quite large. Since 1992 a number of satellites have been recording the change in sea level; they display an acceleration in the rate of sea level change, but they have not been operating for long enough to work out whether this is a real signal, or just an artefact of short-term variation.

"ENVIRONMENTAL" IN THE DISASTER CONTEXT

The environment is often seen as the agent/cause of a disaster or perhaps as the carrier. In an earthquake or a flood, for example, the "environment" behaves in ways that bring harm to the communities affected by them—one suddenly finds the environment sitting in one's living room. However, people make choices—farming practices, use and procurement of fuels, selection of building materials and sites, etc.—that significantly affect their vulnerability to environmental disasters. This view mirrors the idea that disaster is a social construct formed by the interaction of human development with natural processes. An earthquake is a disaster only when it impacts the human infrastructure. But the environment also interacts with human society in complex ways. Floods may damage natural habitats and ecosystems; forest fires may harm forest ecosystems and damage the biotic stock in an area. Yet, floods are necessary to renew and enrich riparian corridors and wetlands and to recharge aquifers; forest fires thin out undergrowth that could fuel larger fires, and they can re-vitalize biodiversity (Sauri 2004). Floods can clog wastewater treatment plants, causing the release of untreated sewage into water bodies; floods can also mobilize contaminants and industrial chemicals that then flow downstream and possibly into those same aquifers.

Thus, an "environmental" hazard may be difficult to define, and there can be a fine distinction between an environmental hazard (*i.e.*, water out of control—a flood) and an environmental resource (*i.e.,* water in control—a reservoir). It can often be a matter of perception regarding deviations about the norm—too much rain is a flood; too little is a drought. Some definitions of environmental hazard emphasize the acute and short-term at the expense of the chronic and long-term (droughts desertification, erosion), for example:

- ...extreme geophysical events, biological processes and major technological accidents, characterized by concentrated releases of energy or materials, which pose a largely unexpected threat to human life and can cause significant damage to goods and the environment

There is a growing understanding of environmental degradation as a contributing factor in disaster effects—*i.e.,* an exacerbating factor in damage, it worsens impact on victims and makes recovery more difficult. One example: Although the largest danger facing Turkish urban areas is earthquake, numerous other hazards exist. Improper handling of solid wastes causes explosive methane build-up, endangers the physical environment, reduces property values and destroys the scenic and tourist values of highly visited areas.... Near the larger cities, many bodies of water are so polluted that they are no longer suitable for recreational use. High levels of heavy metals are found in harbour catches, and massive fish kills are common. Marine accidents release massive, toxic discharges, sometimes causing explosions that destroy buildings and facilities. Dangerous chemicals enter the urban food chain...urban rivers are polluted...agricultural chemicals and waste water have contaminated precious aquifers.

The most recent example occurred in the South Asian tsunami—long-term damage to coral reefs and degradation of mangrove swamps in some areas reduced the capacity of natural systems to absorb or cushion the kinetic energy of the tsunami surge. Deleterious effects of degraded environmental conditions are felt most keenly (though not exclusively) by the poor, residents of shantytowns, "favelas," and other marginal or hazardous areas. They are clustered on steep slopes subject to flash floods and erosion, in dwellings built of substandard materials, with poor water and waste disposal systems. Natural disaster effects can be greatly magnified by the poor environment in which these people live.

According to Pelling (2003b), there is a tendency to focus on technical and engineering issues in addressing environmental problems or issues and to discount the influence of social characteristics on susceptibility to environmental risk. This bias towards technological and physical solutions (*e.g.,* flood walls, or leachate mitigation systems) can encourage development in hazard areas when, in fact, hazards can surpass the margin of safety provided by technological solutions.

"Disaster" in the Environmental Context

The field of emergency management tends to focus more on harm to the human environment and the built environment and to pay less attention to the larger environment in which humans and structures exist. Also, the emphasis is on the more acute disasters (like earthquakes or chemical spills) and less on the slow-developing problems with chronic effects (*e.g.*, Minamata or acid rain) or on acute events with long-lasting consequences (*e.g.*, Bhopal, or the Tisza River). This no doubt reflects the understandable orientation of emergency management professionals to the needs of planning for and response to the immediate effects of a disaster and the desire for speedy restoration to something approaching the status quo ante. Environmental professionals take a somewhat more comprehensive view, considering not only the human and built environments but also the matrix in which they exist. Environmental concerns include not only humans but also plants and animals, water and air quality, the fate and transport of environmental contaminants, the toxicology of human and animal effects, and the exposure and vulnerability (both acute and chronic) of the affected biota.

The Regulatory Imperative

Starting with the National Environmental Policy Act, 28 major environmental protection laws were enacted between 1969 and 1986.

Environmental legislation since 1986 has generally focused on expanding or extending (and, in some cases, clarifying) existing laws. A number of these laws—and their implementing regulations—specifically address emergency planning and/or response in some way.

These include:

- Resource Conservation and Recovery Act (RCRA, 1976)—Requires hazardous waste facilities to prepare and maintain emergency plans to prevent or respond to releases of hazardous wastes
- Comprehensive Environmental Response Compensation and Liability Act (CERCLA or "Superfund")—includes requirements for emergency plans during cleanup actions on uncontrolled waste sites
- Clean Water Act & Oil Pollution Act (1990)—requires Spill Prevention, Control, and Countermeasures plans be prepared by certain facilities storing petroleum fuels
- Emergency Planning & Community Right-to-Know Act (EPCRA or SARA Title III)—directed states and local governments to establish planning and coordination bodies to carry out emergency planning for chemical emergencies in their jurisdictions
- Clean Air Act, section 112r, Risk Management Programme—requires

certain industrial facilities to prepare an "off-site consequence analysis" for releases of certain chemicals and to prepare emergency plans in coordination with local response agencies

- Executive Order 12856 (1993)—directs Federal facilities to comply with EPCRA regarding public notification of chemical use and emergency planning

States have enacted their own set of environmental laws and regulations that parallel, enhance, and extend Federal regulations. In many cases (California and New Jersey are good examples), state regulations are more strict than Federal requirements. An example of European regulatory action, affecting more than one country, is the Seveso Directive. A 1976 explosion at a chemical plant in a small town near Milan, Italy released a large cloud of dioxin that affected a large portion of Lombardy region. The explosion and aftermath, including the botched response, led to creation of the European Community's Seveso Directive in 1982.

A central part of the Directive is a requirement for public information about major industrial hazards and appropriate safety measures in the event of an accident. It is based on recognition that industrial workers and the general public need to know about hazards that threaten them and about safety procedures. This is the first time that the principle of "need to know" has been enshrined in European Community legislation. Much of the Seveso Directive is analogous to EPCRA with additional elements addressing what are called, in the US, "worker Right-To-Know" laws. All of this happened several years prior to the Bhopal disaster, which was part of the impetus for passage of EPCRA in the United States.

Though not regulatory in nature, there are a number of international standards regarding environmental management that specify a requirement for emergency plans. The most widely recognized is the ISO 14001 standard for Environmental Management Systems, one element of which states, "The organization shall establish and maintain procedures to identify potential for and respond to accidents and emergency situations, and for preventing and mitigating the environmental impacts that may be associated with them".

The objective of the Guiding Principles for Chemical Accident Prevention, Preparedness and Response (2003), published by the European Organisation for Economic Co-Operation and Development is to "provide guidance, applicable worldwide ... to prevent accidents involving hazardous substances and to mitigate the adverse effects of accidents that do nevertheless occur." This set of principles covers much the same ground as the EPCRA, the Risk Management programme regulations, hazardous materials transportation regulations, and various environmental and safety standards in the US.

The nexus of Environmental Management, Development, and Disaster Risk

Considerable research and analysis has been done by the European Union and the United Nations to illuminate the connections among environmental hazards, sustainable development strategies (especially in the poorer countries), and disaster response and management. Living with Risk (2004), produced by the UN International Strategy for Disaster Reduction, puts it most succinctly: The environment and disasters are inherently linked. Environmental degradation affects natural processes, alters humanity's resource base and increases vulnerability. It exacerbates the impact of natural hazards, lessens overall resilience and challenges traditional coping strategies. Furthermore, effective and economical solutions to reduce risk can be overlooked.... Although the links between disaster reduction and environmental management are recognized, little research and policy work has been undertaken on the subject. The concept of using environmental tools for disaster reduction has not yet been widely applied by practitioners. The UN International Strategy for Disaster Reduction also focuses on the transboundary nature of disasters and the importance of a "harmonized approach" to the management of pollution of river basins, seismic hazard areas, and volcanoes. This issue is perhaps less salient in the United States, due to the extent of Federal disaster management and response. Researchers in the Swedish Embassy in Bangkok have sought to link environmental programmes with disaster risk in the context of sustainable development.

They ask:

- How can investments in environmental management and sustainable development also reduce disaster risk?
- Is there a prevention dividend that accrues from wise land use planning and development programmes?
- Can prevention dividends be measured; and, how might the ability to estimate these added values enhance policy and programme planning?

Although they find evidence for positive answers to these questions, they acknowledge that more research and analysis is necessary in order to capture the rather elusive cost/benefit parameters of disaster reduction and sustainable development.

Zones of Convergence

Living with Risk (2004, p. 303) outlines ways to integrate environmental and disaster reduction strategies:

- Assessment of environmental causes of hazards occurrence and vulnerability

- Assessment of environmental actions that can reduce vulnerability
- Assessment of the environmental consequences of disaster reduction actions
- Consideration of environmental services in decision-making processes
- Partnerships and regional approaches to land use and nature conservation
- Reasonable alternatives to conflicts concerning alternative uses of resources
- Advice and information to involve actors in enhancing the quality of the environment.

Within this context, there are a number of areas where environmental management and emergency management can and should interact more positively for mutual benefit and support. Both fields would benefit from continuing and supporting the current movement in the disaster community from "reactive" disaster response to active risk management and from iterative recovery to pro-active mitigation and prevention. Parallel efforts would transition the environmental field from contaminant clean-up to risk reduction and pollution prevention, from discrete issues management to environmental management systems, and from flood control to floodplain management.

Integration of sustainability considerations into disaster mitigation and recovery can exploit the considerable overlap between environmental management and disaster management. Planners and practitioners in both fields must recognize that the overall objectives of these fields implicitly promote sustainable communities. Sustainability should be considered both prospectively (in sustainable development planning and mitigation) and retrospectively (in response and recovery).

This integration would incorporate and enhance current trends towards "holistic disaster recovery" (also "sustainable recovery") that emphasize betterment of the entire community, including environmental improvement and enhancement, through the recovery process. Living with Risk (2004) is even more direct: Disaster reduction specialists should be encouraged to anticipate environmental requirements under applicable laws and to design projects that address these requirements, coordinating closely with environmental institutions. Environmental management professionals can make considerable contributions during the mitigation and recovery phases of emergency management. They can identify possible improvements and enhancements as well as things to avoid. More importantly, after enhancements or improvements are in place, they can monitor and assess environmental performance indicators to ensure that goals are met.

Environmental assessments should be integrated into emergency planning processes, following the Environmental Impact Statement model mandated by the National Environmental Protection Act. Environmental Impact Statements should (but currently do not) specifically include disaster-hazard considerations. Rapid environmental assessments should be conducted as part of disaster damage assessment and should be an integral part of response/ recovery considerations. Both environmental managers and emergency managers must be cognizant of the importance of environmental justice/ equity issues in the context of hazard and vulnerability.

Hazards of any type have a disproportionate impact on the poor and disadvantaged. A number of thorny equity issues are coming to a head in the environmental management world, among them: industrial plant and landfill siting; development in industrial or depressed areas; residential settlement on slopes or in other marginal areas; higher population density; immigrants and language differences; differential access to social services and information sources. Most of these issues have not yet been adequately addressed in emergency management planning or community dialogue.

The United States (and, to a certain extent, other nations) has become sensitized to the possibility that terrorists might attack with Weapons of Mass Destruction (nerve agents, bioweapons, "dirty bombs"). The unpleasant reality is that terrorists don't have to try that hard to create death and destruction. The ubiquitous gasoline tanker would make a handy (and easily procured) bomb; there are over 20,000 chemical plants in the US that contain enough extremely hazardous materials to require reporting under EPCRA. The existence and availability of these and other so-called "weapons of convenience" will require a much closer and more explicit cooperation between environmental professionals and emergency managers to: assess the immediate and long-term threats; to identify both mitigation and response strategies; and to manage long-term recovery and clean-up operations.

5

Environmental Science

Environmental science is an interdisciplinary academic field that integrates physical and biological sciences, (including physics, chemistry, biology, soil science, geology, and geography) to the study of the environment, and the solution of environmental problems.

Environmental science provides an integrated, quantitative, and interdisciplinary approach to the study of environmental systems. Related areas of study include environmental studies and environmental engineering. Environmental studies incor- porate more of the social sciences for understanding human relationships, perceptions and policies towards the environment.

Environmental engineering focuses on design and technology for improving environmental quality.

Environmental scientists work on subjects like the understanding of earth processes, evaluating alternative energy systems, pollution control and mitigation, natural resource management, and the effects of global climate change. Environmental issues almost always include an interaction of physical, chemical, and biological processes. Environmental scientists bring a systems approach to the analysis of environmental problems. Key elements of an effective environmental scientist include the ability to relate space, and time relationships as well as quantitative analysis.

Environmental science came alive as a substantive, active field of scientific investigation in the 1960s and 1970s driven by:

- The need for a multi-disciplinary approach to analyse complex environmental problems,
- The arrival of substantive environmental laws requiring specific environmental protocols of investigation,
- The growing public awareness of a need for action in addressing environmental problems.

RADIATION FROM AN OVERCAST SKY

When the sky is overcast, neither its brightness nor the total amount of radiation received on a given horizontal area is at all constant, even for the same height of the nun, because the clouds in question may be of any kind from the thinnest cirrus that just dims the sun, to the darkest nimbus that reduces even noonday brilliance to twilight. If, however, the sky is completely overcast by an approximately uniform cloud layer dense enough, but not greatly more than enough, to prevent the position of the sun from showing, then the total radiation from this cloud layer onto a horizontal surface is, on the average, slightly greater than that from a clear sky.

Evidently, too, the amount received of this cloud-transmitted radiation generally must increase with increase of height above sea level of the place of reception. The brightness of the cloud layer is surprisingly close to uniform. It is greatest nearly overhead (just a little way off in the direction of the sun), but still nine-tenths as bright half way to the horizon, and half as bright almost at the horizon.

DISPOSAL OF RADIATION OF SURFACE OF THE EARTH

There are just three things that can happen to radiation incident onto any extended object. It must be reflected, transmitted or absorbed. When the object is extremely small it more or less scatters incident radiation, and radiation that just grazes the boundary of an object suffers still another effect which we call diffraction. However, neither scattering nor diffraction occurs when the object is large and its edges are not involved. They do not, therefore, occur in the case of radiation incident on the surface of the earth.

Here then, the incident radiation is all used up by two processes, reflection and absorption, since there is no transmission—no passage of radiation through the earth and out at the other side. The portion reflected is about 70% for snow-covered regions, and 7% for the rest of the world. The remainder is absorbed, that is, 30% wherever there is snow, and 93% at all other places, both land and water.

That which is reflected is lost except in so far as it is absorbed by the air above. The absorbed portion goes largely to heating the upper layers of the soil or water, but not all of it, since a considerable part is consumed in maintaining evaporation, and a much smaller part in effecting plant growth and development. Another relatively small part merely melts snow and ice without raising their temperature above the freezing point.

The heated surface in turn heats the soil or rock by conduction, but appreciably to a depth of only a few feet. The heating of water extends to a greater depth owing partly to the penetration of the rays to some distance below the surface, and partly to the mixing of the water by wave action. The heated surface also warms the air above it both by direct contact and by radiation. Furthermore, the heated air through convection shares its warmth with other and colder air above; and the heat consumed in evaporation at one place is liberated, that is made sensible or temperature-producing, some other place, usually in mid-air, where condensation occurs, and far away. Practically every bit of this heating of earth, ocean and air, and supply of energy for evaporation, plant growth, ice melting, and what not else, comes from the sun—all directly except about one part in half a million that reaches us after reflection by the full moon and the planets.

A negligibly small amount comes from the fixed stars, enough to keep the average temperature of the out-doors air about two millionths of a degree Fahrenheit higher than it otherwise would be. Finally, another very small amount comes from the heated interior of the earth.

Quantity and Effects of Heat from the Interior of the Earth

If the earth had no atmosphere, and if there were no sun or stars to send us a flood of radiation, the supply of heat from the interior (of which four-fifths, roughly, is from radioactive material) alone would keep up the surface temperatures to about 60° absolute, on the Fahrenheit scale, that is, -400° F., approximately.

Hence the flow of heat from the interior of the earth per square foot of surface is sufficient to raise the temperature of a gallon of water about 1° F. in 16 days, and the total flow through the whole surface out to space sufficient to heat 92,000 tons of water from the freezing to the boiling point every second of time; or enough, starting at room temperatures, to melt 1,000,000 tons of lead per second. These are big figures, and yet all this flow of heat keeps the actual temperature of the surface of the earth only about 1/25 of a degree F. higher than it otherwise would be.

The figures also tell us the surprising story that if 10,000 times as much heat came from the interior of the earth as now actually does come, or, what amounts to the same thing, if everywhere there was a sea of molten cast iron covered over with a layer of rock and dirt only 10 to 12 feet thick, the oceans above could rest thereon serene with no close approach to the boiling point, so excellent an insulator, or poor a conductor, is this material; and that if the dirt and rock crust were 20 feet thick we could go about over it ourselves in perfect comfort!

Outgoing Radiation

On the average, the earth loses to space, or emits to space, by radiation very approximately the same amount of heat each year that it absorbs of incoming radiation during the same time, plus, of course, the supply of heat that reaches the surface from the interior. We know that the loss is substantially equal to the gain because otherwise the surface would be growing warmer from year to year, and we know that this loss is by radiation as there is no other way for the loss to occur—there being no such thing as conduction to empty space.

The amount of this loss, or radiation of the entire earth to space, can be estimated from our knowledge of the incoming radiation and the fraction of it that is ineffective through scattering and reflection, especially by clouds. It can not be measured directly because we have no means of getting out beyond the atmosphere and from that ideal place pointing our heat-gathering apparatus towards the earth.

But, as implied above, we can make a pretty close estimate of the average rate at which radiation is going out from the earth as a whole, and the conclusion is that it is very nearly the same as that from a perfect radiator, or "black body," at the absolute temperature 454° on the Fahrenheit scale, or -6° F. This is sufficient, per square foot of surface, to heat a gallon of water from the freezing to the boiling point in about 20 1/2 hours.

The radiation from the surface of the earth often is very much greater than this value, even twice as great, or more, because the temperature of the surface frequently is far higher than the -6° F., here assumed. On the other hand, at times and places, owing to very low temperatures, it is much less. On the average, however, the radiation from the surface is much in excess of that which finally gets away to space-greater by the amount of return radiation it absorbs (nearly all of it) from clouds and the atmosphere. The surface of the earth radiates at a relatively high temperature, hence in comparative abundance. Some of this radiation goes directly through the atmosphere, but ordinarily most of it is absorbed by the water vapour and clouds in the lower air, and a little by other things, especially ozone (when the sky is clear, for it is above all clouds) and carbon dioxide. That which is absorbed in the lowest layers is, in general, reradiated, but at a lower temperature than that of the surface. This reradiation is in every direction, half of it downward, some of which is absorbed on the way, and the rest by the surface whose initial temperature and radiation it thus helps to maintain; and half upwards to the next higher layers; and so on up and up from layer to layer, but always with decreasing absorption by the air still above and increasing absorption by that below until the entire atmosphere is left behind.

Clouds Do not Check Radiation

It is a well-known fact that during still clear nights the surface of the earth, and, through it, the adjacent air, cool to a much lower temperature, especially over level land and in valleys and bowl-like depressions, than they do when either the sky is clouded, or the wind is strong.

Furthermore, the lower the clouds, other things being equal, the less the cooling. This interesting and important fact is clearly "explained" in many elementary books and numerous articles on the assumption that clouds and winds check the radiation of the surface of the earth, that is, make it radiate more slowly.

That seems very simple, and would be but for one little fly in the ointment—there isn't a word of truth in it. It might do perhaps as a dose of mental paregoric for a kid with the quizz colic, but it is no good for anything else. Just one thing alone ever reduces radiation, and that is decrease of temperature. No! Clouds do not check in the least radiation from the surface below.

They are, however, themselves good radiators; and as their temperature, when they are low, is nearly that of the earth, they send down to it almost as much radiation as it itself emits, and as practically all this cloud radiation is absorbed by the earth, it follows that the surface temperature remains substantially constant. It isn't that the radiation from the earth is checked in the least, but that it receives from the cloud canopy and absorbs wellnigh as much as it itself gives out. Neither is the approximately constant temperature maintained by an appreciable wind owing to any check whatever exerted by it on surface radiation, but to the fact that the net loss of heat thus sustained, and on clear nights it is considerable, is distributed by turbulence through such a deep layer and great quantity of air that the fall in temperature is small even when the total loss of heat is large.

Temperature of Surface Air

When we talk about the "surface" air it often is advisable to explain just what air we have in mind, for this term is quite flexible. We might mean only that air which is in actual molecular contact with the surface, or that which at most is within a few inches of it, or finally, all below the height of eight or ten feet, the air to which we chiefly are exposed while outdoors.

The temperature of the surface air, in any one of these senses, is determined mainly by that of the surface itself. Whatever the temperature of the surface, that also is the temperature of the contact air, and very nearly the temperature of all that air which by turbulence or otherwise is frequently brought into contact with the surface.

On the stillest of nights this layer at places may be only a few thick. During the daytime, however, especially when there is sunshine to induce thermal convection, and whenever, day or night, there is a measurable movement of the air, it is certain to be at least a good many feet thick. The essential point in this: The temperature of the air near the surface (how near varies with the circumstances) depends more on contacts with that surface than it does on the amount of solar radiation to which it may be exposed.

The surface temperature of course does vary with the intensity and duration of the sunshine, and so therefore does also that of the surface air, but indirectly through contact with the surface and not directly by absorption of solar energy.

Relation between Surface Temperature and Temperature of Surface Air

The temperature of the actual contact air must be the same as that of the surface (of the substance, ground or what not, at its surface) against which it rests. If this surface air remained fixed, as we often are told that it does, then it would seem that the air next in contact with it also should become fixed in position, and so on indefinitely.

But we know that fixity of position of the air molecules does not extend to a measurable distance from any solid, for we can blow smoke past it and see the motion of the air. We therefore are forced to the conclusion that fixity of position does not apply, at least not for any appreciable length of time, even to the contact molecules.

The actual contact molecules of the air are at rest, like a liquid film, but they do not stay at rest. They evaporate, and as they leave the surface others condense thereon—are adsorbed—a continuous process the details of which are not yet all known. In this way the contact molecules, during the extremely brief interval of their contact, are fixed in position, but they are continuously reverting to the gaseous state, and therefore the atmosphere at ordinary temperatures is always fluid however measurably near it may be to the surface in question.

Since the air is directly heated chiefly by contact with the surface of the earth, and indirectly by the sharing of this heat, through convection, with colder air above, it follows that in general wherever the temperature of the lower atmosphere is increasing, that is, over nearly all snow-free land, and particularly during the day time, there the average temperature of the surface is higher than that of the free surface air. This is in accordance with what physicists call the second law of thermodynamics, and what everybody else knows without calling it anything, namely, that the temperature of the heater is higher than the temperature of the thing heated. Similarly, where the lower air commonly is cooled, as it is over snow-covered regions, there the average temperature of the surface is lower than that of the surface air—the cooler is colder than the thing cooled.

Maximum and Minimum Temperatures

Obviously if the heater changes temperature, so also will the heated, and the heater will be the first to change and the first to reach its extreme values—maxima and minima. There is no surprise, therefore, in the fact that the daily maximum temperature of a snow-free land surface occurs earlier, about 1 o'clock P.M., than that of the air above it, which is delayed until around 3 o'clock.

The air and surface minima, occurring near daybreak, are much closer together, owing partly to the slow cooling of the soil through the night. Over the ocean the temperature of the air normally is a little higher, a degree or so, than that of the water, and the time of its maximum value, near 1 o'clock P.M., a little earlier than that of the water. This is due to the fact that here the surface air is humid and also "dusty" with salt particles and therefore absorbs a large amount of radiation, so much indeed that its daily range of temperature is more dependent on this direct absorption than it is on conduction and convection from the surface. The minimum temperatures of air and water occur simultaneously, or nearly so.

Periodic Temperature Changes

Nearly 150 different periods of temperature and other weather changes ranging from 24 hours to 744 years. Nearly all of them, however, have a shorter period than 40 years, and half of them a period of 8 years or less. Of this great number of periods there are only two, the 24-hour or daily period, and the 12-month or annual period, that everybody accepts. There is one other, the so-called 11-year or sunspot period, that is widely, though not universally accepted; and still another the 35-year, or Brückner, period that many believe to be real.

No credence was ever given to any of the others save perhaps by their discoverers, and in most cases even that must have been half-hearted. The daily period is everywhere conspicuous (save for part of the time in polar regions, when the sun is continuously above or continuously below the horizon), and in respect to temperature, gives, on the average, a maximum in the early to mid afternoon and a minimum shortly before sunrise.

Over the oceans this diurnal range is only 1° F. to 3° F. as a rule. It also is small in the humid and cloudy portions of the continental tropics, owing to the large amount of return radiation from the clouds and water vapour. In desert regions, especially at high altitudes, where the sky is clear and the humidity very low the diurnal range of temperature is at its maximum. In extreme cases this range is of the order of 100° F., from distinctly below freezing to decidedly over 100° F.—both in the shade. The annual range also is extremely conspicuous in most parts of the world.

In this case the exception does not occur at and near the poles, but at and for some distance on either side of the equator. At the equator the "year," as it were, counted from the time the sun is overhead at noon until farthest away (23 1/2°), and then back again is 6 months, not 12. Next beyond the equator on either side there obviously are two such "years" but of unequal length. At one distance they are 5 and 7 months, at another 4 and 8, and so on until at the Tropics, Capricorn and Cancer, only one is left, and that one 12 months in duration, the same as from there on to the pole.

The times of occurrence of the annual maxima and minima vary widely from place to place, but always they are after maximum and minimum reception of heat from the sun. The delays are least over inland deserts and greatest over mid to high latitude portions of the oceans. The sunspot period, approximately 11.1 years, is most pronounced at high levels within the tropics. Here the average temperature during the year or two around sunspot minima, or when the spots are fewest and smallest, is about 2° F. higher than the average temperature during the time of spot maxima. The same relation appears to hold, in general, for the middle and higher latitudes but with decidedly less contrast.

The Brückner period is very irregular in length, varying from roughly 20 years to perhaps 50, and the amplitude of its temperature range uncertain but always small. Its irregularity in length deprives it of practically all forecasting value, and indeed makes its very existence as anything other than a fortuitous recurrence highly doubtful. There are two other known and real periods in respect to average temperatures and other climatic elements, but they are far too long to consider in any business affairs.

One concerns the slow change of the season of the year when the earth is nearest the sun, due to the combined effect of the motion of the perihelion and the precession of the equinoxes. Just at present the earth is nearest the sun the first week of January and farthest away the first week of July; and this difference in distance is sufficient, if long continued, to vary the average temperature of the earth by at least 7° or 8° F. That is, at present the winters of the northern hemisphere are shorter and milder, and the summers longer and less hot, than they would be if we were nearest the sun the first week of July and farthest from it the first week of January, as we were about 10,500 years ago, and, in the same length of time, will be again.

This is one reason, and the unequal distribution of land and water another, why the average temperature for the year is about 2° F. higher in the northern hemisphere than in the southern, and why the thermal equator is north of the geographic equator.

Beyond question this particular period is of great climatic importance, but we know all about its course and its cause and the changes it effects come about so slowly that practically they do not concern us at all. The other period referred to is that of the changes in the ellipticity of the earth's orbit or variations in the difference between the annual maximum and minimum distances of the earth from the sun. But the length of this period, roughly 100,000 years, keeps it out of every business equation however prudently constructed.

Just to make the list complete it may be worthwhile to mention a few utterly unimportant but entirely real temperature periods. Those are the periods of the changes in light and heat received from the moon-maximum at full moon, minimum at new moon; changes in the distance of the earth from the sun due to the pull of the moon in its orbit about the earth; and similar but far less changes of and by the planets. The sum total of the effects of all the planets is about equal to that of the moon alone, that is, a change in the average temperature of the earth of about.02° F., due almost wholly to variations in our distance from the sun, or, as we say, to perturbations in the earth's orbit. But, as already stated, this change in temperature is too small to bother about.

Temperature Lag

It was stated that the hottest time of the day is not noon, when the sun is most effective, but two to four hours later; and similarly, that the coldest weather does not come with the shortest days, but generally a month or so later. In proverb form: "As the days grow longer the cold grows stronger." In the early morning of a clear day following a cloudless night, say, the earth and surface air are relatively cool.

Then with sunrise they begin to warm up, but not rapidly, even when there is no wind, because it requires an appreciable amount of heat to warm even a pound of soil 1° F., and several times as much to equally warm a pound of water. But as the sunshine continues, the soil at first gets hotter and hotter, and as its temperature rises the rate at which it loses heat by radiation rapidly increases.

However, since the soil, including of course its covering, warms slowly, owing to its large capacity for heat, its loss by radiation falls more and more behind its gain by absorption as the sun rises higher in the heavens, and therefore catches up with the latter only in the afternoon when the insolation is distinctly less than it was at midday.

Hence the diurnal maximum temperature, whether of the lower air or of the surface of the earth (an earlier phenomenon) necessarily lags behind the maximum intensity of the sunshine. Similarly, the annual maximum temperature occurs several weeks after the days are longest and the heating strongest. Very similarly too, because the earth can give off stored up heat when the supply becomes deficient, the minimum temperature comes several weeks after the shortest days. During this period, as the days grow longer the cold grows stronger.

The diurnal and annual heating and cooling, and lagging of temperature extremes, may be likened to the alternate rise and fall of the water level in a reservoir having a continuously open drain pipe at the bottom and a periodically variable inflow, now greater, now less, than the then rate of outflow, but so regulated that the reservoir may never become empty.

Day Degrees

Not only are we interested in the values and times of occurrence of maximum and minimum temperatures but also, and even more, concerned in the occurrence of certain critical temperatures. For instance, we are very much interested in the temperature at which frost can occur until it does occur, after which, if it has been a "killing" one, we are no longer much concerned as there is nothing left for the next one to injure. Another critical temperature is 42° F. as that closely marks the boundary between growth and dormancy for most vegetation of the temperate zones.

In fact it is customary to call the difference between the average temperature of a given day, if higher than this value, and 42° F., its day degrees. The sum of these daily values over a week, month or season, is the number of day degrees for that period, and is an important index to what might have been the vegetable growth during the time in question. Similarly, engineers and others interested in artificial heating of buildings, count day degrees relative to a temperature of 65° F.

Occasional Extremes

Once in a while an exceptional combination of conditions brings to a given place an abnormally high or low temperature, usually for only an hour or two, or a day at most, but sometimes for several days together, and even a month or longer. It is always easy to know from the current maps of weather distribution exactly what caused the extreme in question, but it never is possible to trace them farther back than two or three steps at most, nor very long to foresee their coming. Some of them one never forgets, and a few continue for a century or more to put disconcerting humps or depressions on our statistical curves.

Wind Direction and Temperature

The effect of wind direction on the temperature of a place depends on its location. Well within the Tropics, and also near the poles, the effect of wind direction obviously is small because the temperature is pretty nearly the same round about in every direction. In middle latitudes, however, the situation is quite different, partly because here the temperatures commonly are not the same in every direction, and partly also, in fact mainly, because here each section of the cyclone and of the anticyclone has its own wind direction, and some of them a wind system entirely distinct from that of the others.

In the forward or eastern portion of the anticyclone the winds are from the region of higher latitudes, and having come a long ways often are distinctly cool to cold for the place and time of year. Similarly, the winds of the western segment, having come from much nearer the equator, usually are relatively warm. In the cyclone, or widespread disturbance, all that segment of 90°, more or less, lying between one line running east, to southeast, from the storm centre and another generally south to southwest (in the northern hemisphere; east to northeast, and north to northwest, in the southern hemisphere) is occupied by a great current of warm air from low latitudes.

The rest of the storm area is covered with cold winds from the east, north, and northwest, in succession as one in the northern hemisphere passes from the front to the rear of the storm centre on the poleward side; from the east, south, and southwest, in the southern hemisphere. In general, all these cold winds in each hemisphere are of polar, that is, high latitude origin.

Clearly then, the temperature of the air in a cyclonic region is likely to change with the direction of the wind. In one portion of this disturbance, namely, along a narrow strip that meteorologists call the cold front, or wind shift line, and which commonly runs west of south (west of north in the southern hemisphere) from the storm centre, the wind direction rapidly changes from southwesterly to northwesterly, with, as a rule, a sharp drop in temperature as the tropical breezes give way to polar blasts. Hence in middle latitudes air temperature is closely dependent upon wind direction, both in cyclones and anticyclones; and that means the greater portion of the time, for usually we are in the midst of one or the other of these disturbances.

METHODS OF GENERATING ELECTRICITY

Turbines

Rotating turbines attached to electrical generators produce most commercially available electricity. Turbines are driven by a fluid which acts as an intermediate energy carrier. The fluids typically used are:

- *Steam:* Water is boiled by nuclear fission or the burning of fossil fuels (coal, natural gas, or petroleum). Some newer plants use the sun as the heat source: solar parabolic troughs and solar power towers concentrate sunlight to heat a heat transfer fluid, which is then used to produce steam.

- *Water:* Turbine blades are acted upon by flowing water, produced by hydroelectric dams or tidal forces,
- *Wind:* Most wind turbines generate electricity from naturally occurring wind. Solar updraft towers use wind that is artificially produced inside the chimney by heating it with sunlight.
- *Hot Gases:* Turbines are driven directly by gases produced by the combustion of natural gas or oil.

Combined cycle gas turbine plants are driven by both steam and gas. They generate power by burning natural gas in a gas turbine and use residual heat to generate additional electricity from steam. These plants offer efficiencies of up to 60%.

Reciprocating Engines

Small electricity generators are often powered by reciprocating engines burning diesel, biogas or natural gas. Diesel engines are often used for back up generation, usually at low voltages. Biogas is often combusted where it is produced, such as a landfill or wastewater treatment plant, with a reciprocating engine or a microturbine, which is a small gas turbine.

Photovoltaic Panels

Unlike the solar heat concentrators mentioned above, photovoltaic panels convert sunlight directly to electricity. Although sunlight is free and abundant, solar panels are expensive to produce and have only a 10-20% conversion efficiency. Until recently, photovoltaics were most commonly used in remote sites where there is no access to a commercial power grid, or as a supplemental electricity source for individual homes and businesses. Recent advances in manufacturing efficiency and photovoltaic technology, combined with subsidies driven by environmental concerns, have dramatically accelerated the deployment of solar panels. Installed solar capacity is growing by 30% per year in several regions including Germany, Japan, California and New Jersey.

Nuclear Power

A nuclear power station. Condensed water vapor rises from the hyperboloid shaped cooling towers. The nuclear reactors are inside the cylindrical containment buildings.

Nuclear power is the controlled use of nuclear reactions to release energy for work including propulsion, heat, and the generation of electricity. Human use of nuclear power to do significant useful work is currently limited to nuclear fission and radioactive decay. Nuclear energy is produced when a fissile material, such as uranium-235 (235U), is concentrated such that nuclear fission takes place in a controlled chain reaction and creates heat-which is used to boil water, produce steam, and drive a steam turbine. The turbine can be used for mechanical work and also to generate electricity. Nuclear power is used to power most military submarines and aircraft carriers and provides 7% of the world's energy and 15.7% of the world's electricity.

The United States produces the most nuclear energy, with nuclear power providing 20% of the electricity it consumes, while France produces the highest percentage of its electrical energy from nuclear reactors-80% as of 2006. Nuclear energy policy differs between countries.

Nuclear energy uses an abundant, widely distributed fuel, and mitigates the greenhouse effect if used to replace fossil-fuel-derived electricity. International research is ongoing into various safety improvements, the use of nuclear fusion and additional uses such as the generation of hydrogen (in support of hydrogen economy schemes), for desalinating sea water, and for use in district heating systems.

Construction of nuclear power plants in the U.S. declined following the 1979 Three Mile Island accident and the 1986 disaster at Chernobyl. Lately, there has been renewed interest in nuclear energy from national governments due to economic and environmental concerns. Other reasons for interest include the public, some notable environmentalists due to increased oil prices, new passively safe designs of plants, and the low emission rate of greenhouse gas which some governments need to meet the standards of the Kyoto Protocol. A few reactors are under construction, and several new types of reactors are planned.

The use of nuclear power is controversial because of the problem of storing radioactive waste for indefinite periods, the potential for possibly severe radioactive contamination by accident or sabotage, and the possibility that its use in some countries could lead to the proliferation of nuclear weapons. Proponents believe that these risks are small and can be further reduced by the technology in the new reactors. They further claim that the safety record is already good when compared to other fossil-fuel plants, that it releases much less radioactive waste than coal power, and that nuclear power is a sustainable energy source. Critics, including most major environmental groups, believe nuclear power is an uneconomic, unsound and potentially dangerous energy source, especially compared to renewable energy, and dispute whether the costs and risks can be reduced through new technology. There is concern in some countries over North Korea and Iran operating research reactors and fuel enrichment plants, since those countries refuse adequate IAEA oversight and are believed to be trying to develop nuclear weapons. North Korea admits that it is developing nuclear weapons, while the Iranian government vehemently denies the claims against Iran.

Origins

The first successful experiment with nuclear fission was conducted in 1938 in Berlin by the German physicists Otto Hahn, Lise Meitner and Fritz Strassmann.

During the Second World War, a number of nations embarked on crash programs to develop nuclear energy, focusing first on the development of nuclear reactors. The first self-sustaining nuclear chain reaction was obtained at the University of Chicago by Enrico Fermi on December 2, 1942, and reactors based on his research were used to produce the plutonium necessary for the "Fat Man" weapon dropped on Nagasaki, Japan. Several nations began their own construction of nuclear reactors at this point, primarily for weapons use, though research was also being conducted into their use for civilian electricity generation.

Electricity was generated for the first time by a nuclear reactor on December 20, 1951 at the EBR-I experimental fast breeder station near Arco, Idaho, which initially produced about 100 kW.

In 1952 a report by the Paley Commission (The President's Materials Policy Commission) for President Harry Truman made a "relatively pessimistic" assessment of nuclear power, and called for "aggressive research in the whole field of solar energy".

A December 1953 speech by President Dwight Eisenhower, "Atoms for Peace", set the U.S. on a course of strong government support for the international use of nuclear power.

Early Years

The Beaver Valley Nuclear Generating Station in Shippingport, Pennsylvania was the first commercial reactor in the USA and was opened in 1957.

On June 27, 1954, the world's first nuclear power plant to generate electricity for a power grid started operations at Obninsk, USSR. The reactor was graphite moderated, water cooled and had a capacity of 5 megawatts (MW). The world's first commercial nuclear power station, Calder Hall in Sellafield, England was opened in 1956, a gas-cooled Magnox reactor with an initial capacity of 50 MW (later 200 MW). The Shippingport Reactor (Pennsylvania, 1957), a pressurized water reactor, was the first commercial nuclear generator to become operational in the United States. In 1954, the chairman of the United States Atomic Energy Commission (forerunner of the U.S. Nuclear Regulatory Commission) talked about electricity being "too cheap to meter" in the future, often misreported as a concrete statement about nuclear power, and foresaw 1000 nuclear plants on line in the USA by the year 2000.

In 1955 the United Nations' "First Geneva Conference", then the world's largest gathering of scientists and engineers, met to explore the technology. In 1957 EURATOM was launched alongside the European Economic Community (the latter is now the European Union). The same year also saw the launch of the International Atomic Energy Agency (IAEA).

Development

Installed nuclear capacity initially rose relatively quickly, rising from less than 1 gigawatt (GW) in 1960 to 100 GW in the late 1970s, and 300 GW in the late 1980s. Since the late 1980s capacity has risen much more slowly, reaching 366 GW in 2005, primarily due to Chinese expansion of nuclear power. Between around 1970 and 1990, more than 50 GW of capacity was under construction (peaking at over 150 GW in the late 70s and early 80s)- in 2005, around 25 GW of new capacity was planned. More than two-thirds of all nuclear plants ordered after January 1970 were eventually cancelled.

During the 1970s and 1980s rising economic costs (related to vastly extended construction times largely due to regulatory delays) and falling fossil fuel prices made nuclear power plants then under construction less attractive. In the 1980s (U.S.) and 1990s (Europe), flat load growth and electricity liberalization also made the addition of large new baseload capacity unnecessary.

A general movement against nuclear power arose during the last third of the 20th century, based on the fear of a possible nuclear accident and on fears of latent radiation, and on the opposition to nuclear waste production, transport and final storage.

Perceived risks on the citizens' health and safety, the 1979 accident at Three Mile Island and the 1986 Chernobyl accident played a key part in stopping new plant construction in many countries. Austria (1978), Sweden (1980) and Italy (1987) voted in referendums to oppose or phase out nuclear power, while opposition in Ireland prevented a nuclear programme there. However, the Brookings Institution suggests that new nuclear units have not been ordered primarily for economic reasons rather than fears of accidents.

Financing for new reactors dried up when Wall Street's enthusiasm ended. Disillusionment was complete when new research discredited the claim (previously accepted as fact even by opponents) that nuclear power was still, despite all its problems, the most cost-effective source of electricity. Industry figures had omitted the factor of downtime. During the 1980s and early 1990s, the newest and biggest U.S. plants were actually producing only half the energy they were supposed to, due to shutdowns for refueling, routine maintenance, retrofitting, and frequent minor mishaps. Since that time, the capacity factor of existing nuclear power plants has increased dramatically, and has been near 90% in the current decade. As of 2006, the stated desire to use nuclear power for electricity generation has been suspected of being a cover for nuclear proliferation in the countries of Iran and North Korea.

Reactor Types

Current Technology: There are two types of nuclear power in current use:

1. The nuclear fission reactor produces heat through a controlled nuclear chain reaction in a critical mass of fissile material.

All current nuclear power plants are critical fission reactors, which are the focus of this article. The output of fission reactors is controllable. There are several subtypes of critical fission reactors, which can be classified as Generation I, Generation II and Generation III. All reactors will be compared to the Pressurized Water Reactor (PWR), as that is the standard modern reactor design.

The difference between fast-spectrum and thermal-spectrum reactors will be covered later. In general, fast-spectrum reactors will produce less waste, and the waste they do produce will have a vastly shorter halflife, but they are more difficult to build, and more expensive to operate. Fast reactors can also be breeders, whereas thermal reactors generally cannot.

Pressurized Water Reactors (PWR)

These are reactors cooled and moderated by high pressure liquid (even at extreme temperatures) water. They are the majority of current reactors, and are generally considered the safest and most reliable technology currently in large scale deployment, although Three Mile Island is a reactor of this type. This is a thermal neutron reactor design.

Boiling Water Reactors (BWR)

These are reactors cooled and moderated by water, under slightly lower pressure. The water is allowed to boil in the reactor. The thermal efficiency of these reactors can be higher, and they can be simpler, and even potentially more stable and safe. Unfortunately, the boiling water puts more stress on many of the components, and increases the risk that radioactive water may escape in an accident. These reactors make up a substantial percentage of modern reactors. This is a thermal neutron reactor design.

Pressurized Heavy Water Reactor (PHWR)

A Canadian design, (known as CANDU) these reactors are heavy-water-cooled and-moderated Pressurized-Water reactors. Instead of using a single large containment vessel as in a PWR, the fuel is contained in hundreds of pressure tubes. These reactors are fuelled with natural uranium and are thermal neutron reactor designs.

PHWRs can be refueled while at full power, which makes them very efficient in their use of uranium (it allows for precise flux control in the core). Most PHWRs exist within Canada, but units have been sold to Argentina, China, India (pre-NPT), Pakistan (pre-NPT), Romania, and South Korea. India also operates a number of PHWR's, often termed 'CANDU-derivatives', built after the 1974 Smiling Buddha nuclear weapon test.

Reaktor Bolshoy Moshchnosti Kanalniy (RBMK)

A Soviet Union design, built to produce plutonium as well as power, the dangerous and unstable RBMKs are water cooled with a graphite moderator. RBMKs are in some respects similar to CANDU in that they are refuelable On-Load and employ a pressure tube design instead of a PWR-style pressure vessel.

However, unlike CANDU they are very unstable and too large to have containment buildings. Because of this RBMK reactors are generally considered one of the most dangerous reactor designs in use. Chernobyl was an RBMK.

Gas Cooled Reactor (GCR) and Advanced Gas Cooled Reactor (AGCR)

These are generally graphite moderated and CO2 cooled. They have a high thermal efficiency compared with PWRs and an excellent safety record.

There are a number of operating reactors of this design, mostly in the United Kingdom. Older designs (i.e. Magnox stations) are either shut down or will be in the near future. However, the AGCRs have an anticipated life of a further 10 to 20 years. This is a thermal neutron reactor design.

Super Critical Water-cooled Reactor (SCWR)

This is a theoretical reactor design that is part of the Gen-IV reactor project. It combines higher efficiency than a GCR with the safety of a PWR, though it is perhaps more technically challenging than either.

The water is pressurized and heated past its critical point, until there is no difference between the liquid and gas states. An SCWR is similar to a BWR, except there is no boiling (as the water is critical), and the thermal efficiency is higher as the water behaves more like a classical gas. This is an epithermal neutron reactor design.

Liquid Metal Fast Breeder Reactor (LMFBR)

This is a reactor design that is cooled by liquid metal, totally unmoderated, and produces more fuel than it consumes. These reactors can function much like a PWR in terms of efficiency, and do not require much high pressure containment, as the liquid metal does not need to be kept at high pressure, even at very high temperatures.

Superphénix in France was a reactor of this type, as was Fermi-I in the United States. The Monju reactor in Japan suffered a sodium leak in 1995 and is approved for restart in 2008. All three use/used liquid sodium. These reactors are fast neutron, not thermal neutron designs. These reactors come in two types:

Lead Cooled

Using lead as the liquid metal provides excellent radiation shielding, and allows for operation at very high temperatures. Also, lead is (mostly) transparent to neutrons, so fewer neutrons are lost in the coolant, and the coolant does not become radioactive. Unlike sodium, lead is mostly inert, so there is less risk of explosion or accident, but such large quantities of lead may be problematic from toxicology and disposal points of view. Often a reactor of this type would use a lead-bismuth eutectic mixture. In this case, the bismuth would present some minor radiation problems, as it is not quite as transparent to neutrons, and can be transmuted to a radioactive isotope more readily than lead.

Sodium Cooled

Most LMFBRs are of this type. The sodium is relatively easy to obtain and work with, and it also manages to actually remove corrosion on the various reactor parts immersed in it. However, sodium explodes violently when exposed to water, so care must be taken, but such explosions wouldn't be vastly more violent than (for example) a leak of superheated fluid from a SCWR or PWR.

The radioisotope thermoelectric generator produces heat through passive radioactive decay.

Some radioisotope thermoelectric generators have been created to power space probes (for example, the Cassini probe), some lighthouses in the former Soviet Union, and some pacemakers. The heat output of these generators diminishes with time; the heat is converted to electricity utilising the thermoelectric effect.

Working of Nuclear Power Plants

The key components common to most types of nuclear power plants are:

- Nuclear fuel
- Neutron moderator
- Coolant

- Control rods
- Pressure vessel
- Emergency core cooling systems
- Reactor protective system
- Steam generators (not in BWRs)
- Containment building
- Boiler feedwater pump
- Turbine
- Electrical generator
- Condenser

Conventional thermal power plants all have a heat source. Examples are gas, coal, or oil. For a nuclear power plant, this heat is provided by nuclear fission inside the nuclear reactor. When a relatively large fissile atomic nucleus (usually uranium-235 or plutonium-239) is struck by a neutron it forms two or more smaller nuclei as fission products, releasing energy and neutrons in a process called nuclear fission.

The neutrons then trigger further fission. And so on. When this nuclear chain reaction is controlled, the energy released can be used to heat water, produce steam and drive a turbine that generates electricity. It should be noted that a nuclear explosive involves an uncontrolled chain reaction, and the rate of fission in a reactor is not capable of reaching sufficient levels to trigger a nuclear explosion because commercial reactor grade nuclear fuel is not enriched to a high enough level.

The chain reaction is controlled through the use of materials that absorb and moderate neutrons. In uranium-fueled reactors, neutrons must be moderated (slowed down) because slow neutrons are more likely to cause fission when colliding with a uranium-235 nucleus. Light water reactors use ordinary water to moderate and cool the reactors. When at operating temperatures if the temperature of the water increases, its density drops, and fewer neutrons passing through it are slowed enough to trigger further reactions. That negative feedback stabilizes the reaction rate.

Experimental Technologies

A number of other designs for nuclear power generation, the Generation IV reactors, are the subject of active research and may be used for practical power generation in the future. A number of the advanced nuclear reactor designs could also make critical fission reactors much cleaner, much safer and/ or much less of a risk to the proliferation of nuclear weapons.

- *Integral Fast Reactor:* The link at the end of this paragraph references an interview with Dr. Charles Till, former director of Argonne National

Laboratory West in Idaho and outlines the Integral Fast Reactor and its advantages over current reactor design, especially in the areas of safety, efficient nuclear fuel usage and reduced waste. The IFR was built, tested and evaluated during the 1980s and then retired under the Clinton administration in the 1990s due to nuclear non-proliferation policies of the administration. Recycling spent fuel is the core of its design and it therefore produces a fraction of the waste of current reactors.

- *Pebble Bed Reactor:* This reactor type is designed so high temperatures reduce power output by doppler broadening of the fuel's neutron cross-section. It uses ceramic fuels so its safe operating temperatures exceed the power-reduction temperature range. Most designs are cooled by inert helium, which cannot have steam explosions, and which does not easily absorb neutrons and become radioactive, or dissolve contaminants that can become radioactive. Typical designs have more layers (up to 7) of passive containment than light water reactors (usually 3). A unique feature that might aid safety is that the fuel-balls actually form the core's mechanism, and are replaced one-by-one as they age. The design of the fuel makes fuel reprocessing expensive.
- SSTAR, Small, Sealed, Transportable, Autonomous Reactor is being primarily researched and developed in the US, intended as a fast breeder reactor that is tamper resistant and passively safe.
- Subcritical reactors are designed to be safer and more stable, but pose a number of engineering and economic difficulties.
- Controlled nuclear fusion could in principle be used in fusion power plants to produce safer, cleaner power, but significant scientific and technical obstacles remain. Several fusion reactors have been built, but as yet none has 'produced' more thermal energy than electrical energy consumed. Despite research having started in the 1950s, no commercial fusion reactor is expected before 2050. The ITER project is currently leading the effort to commercialize fusion power.
- Thorium based reactors. It is possible to convert Thorium-232 into U-233 in reactors specially designed for the purpose. In this way, Thorium, which is more plentiful than uranium, can be used to breed U-233 nuclear fuel. U-233 is also believed to have favourable nuclear properties as compared to traditionally used U-235, including better neutron economy and lower production of long lived transuranic waste.
- *Advanced Heavy Water Reactor:* A proposed heavy water moderated nuclear power reactor that will be the next generation design of the PHWR type. Under development in the Bhabha Atomic Research Centre (BARC).

- KAMINI-A unique reactor using Uranium-233 isotope for fuel. Built by BARC and IGCAR Uses thorium.
- India is also building a bigger scale FBTR or fast breeder thorium reactor to harness the power with the use of thorium.

Life Cycle

The Nuclear Fuel Cycle begins when uranium is mined, enriched, and manufactured into nuclear fuel, (1) which is delivered to a nuclear power plant. After usage in the power plant, the spent fuel is delivered to a reprocessing plant (2) or to a final repository (3) for geological disposition. In reprocessing 95% of spent fuel can be recycled to be returned to usage in a power plant (4).

Nuclear Fuel

A nuclear reactor is only part of the life-cycle for nuclear power. The process starts with mining. Generally, uranium mines are either open-pit strip mines, or in-situ leach mines. In either case, the uranium ore is extracted, usually converted into a stable and compact form such as yellowcake, and then transported to a processing facility. Here, the yellowcake is converted to uranium hexafluoride, which is then enriched using various techniques. At this point, the enriched uranium, containing more than the natural 0.7% U-235, is used to make rods of the proper composition and geometry for the particular reactor that the fuel is destined for. The fuel rods will spend about 3 years inside the reactor, generally until about 3% of their uranium has been fissioned, then they will be moved to a spent fuel pool where the short lived isotopes generated by fission can decay away. After about 5 years in a cooling pond, the spent fuel is radioactively cool enough to handle, and it can be moved to dry storage casks or reprocessed.

Fuel Resources

Uranium is a common element, occurring almost everywhere on land and in the oceans. It is about as common as tin, and 500 times more common than gold. Most types of rocks and soils contain uranium, although often in low concentrations. At present, economically viable deposits are regarded as being those with concentrations of at least 0.1% uranium. At this cost level, available reserves would last for 50 years at the present rate of use. Doubling the price of uranium, which would have only little effect on the overall cost of nuclear power, would increase reserves to hundreds of years. To put this in perspective; a doubling in the cost of natural uranium would increase the total cost of nuclear power by 5%. On the other hand, if the price of natural gas was doubled, the cost of gas-fired power would increase by about 60%.

Doubling the price of coal would increase the cost of power production in a large coal-fired power station by about 30%. Uranium enrichment produces many tons of depleted uranium (DU) which consists of U-238 with most of the easily fissile U-235 isotope removed. U-238 is a tough metal with several commercial uses-for example, aircraft production, radiation shielding, and making bullets and armor-as it has a higher density than lead. There are concerns that U-238 may lead to health problems in groups exposed to this material excessively, like tank crews and civilians living in areas where large quantities of DU ammunition have been used.

Current light water reactors make relatively inefficient use of nuclear fuel, leading to energy waste. More efficient reactor designs or nuclear reprocessing would reduce the amount of waste material generated and allow better use of the available resources.

As opposed to current light water reactors which use uranium-235 (0.7% of all natural uranium), fast breeder reactors use uranium-238 (99.3% of all natural uranium). It has been estimated that there is up to five-billion years' worth of uranium-238 for use in these power plants. Breeder technology has been used in several reactors. Currently (December 2005), the only breeder reactor producing power is BN-600 in Beloyarsk, Russia. (The electricity output of BN-600 is 600 MW-Russia has planned to build another unit, BN-800, at Beloyarsk nuclear power plant.) Also, Japan's Monju reactor is planned for restart (having been shut down since 1995), and both China and India intend to build breeder reactors. Another alternative would be to use uranium-233 bred from thorium as fission fuel-the thorium fuel cycle. Thorium is three times more abundant in the Earth's crust than uranium, and (theoretically) all of it can be used for breeding, making the potential thorium resource orders of magnitude larger than the uranium fuel cycle operated without breeding. Unlike the breeding of U-238 into plutonium, fast breeder reactors are not necessary-it can be performed satisfactorily in more conventional plants. Proposed fusion reactors assume the use of deuterium, an isotope of hydrogen, as fuel and in most current designs also lithium. Assuming a fusion energy output equal to the current global output and that this does not increase in the future, then the known current lithium reserves would last 3,000 years, lithium from sea water would last 60 million years, and a more complicated fusion process using only deuterium from sea water would have fuel for 150 billion years. For comparison, the Sun has an estimated remaining life of 5 billion years.

Solid Waste

The predominant waste stream from nuclear power plants is spent fuel. A large nuclear reactor produces 3 cubic metres (25-30 tonnes) of spent fuel each year. It is primarily composed of unconverted uranium as well as significant quantities of transuranic actinides (plutonium and curium, mostly). In addition, about 3% of it is made of fission products. The actinides (uranium, plutonium, and curium) are responsible for the bulk of the long term radioactivity, whereas the fission products are responsible for the bulk of the short term radioactivity.

Spent fuel is highly radioactive and needs to be handled with great care and forethought. Fresh from the reactor, it is so radioactive that less than a minute's exposure to it will cause death. However, spent nuclear fuel becomes less radioactive over time. After 40 years, the radiation flux is 99.9% lower than it was the moment the spent fuel was removed, although still dangerously radioactive.

The safe storage and disposal of nuclear waste is a significant challenge. Because of potential harm from radiation, spent nuclear fuel must be stored in shielded basins of water (spent fuel pools), and usually subsequently in dry storage vaults or dry cask storage until its radioactivity decreases naturally ("decays") to safe levels. This interim stage spans years or decades, depending on the type of fuel. Most U.S. waste is currently stored in temporary storage sites requiring oversight, while suitable permanent disposal methods are discussed. As of 2003, the United States had accumulated about 49,000 metric tons of spent nuclear fuel from nuclear reactors. Underground storage at Yucca Mountain in U.S. has been proposed as permanent storage. After 10,000 years of radioactive decay, according to United States Environmental Protection Agency standards, the spent nuclear fuel will no longer pose a threat to public health and safety.

The amount of waste can be reduced in several ways, particularly reprocessing, as below. Even so, the remaining waste will be substantially radioactive for at least 300 years even if the actinides are removed, and for up to thousands of years if the actinides are left in. Even with separation of all actinides, and using fast breeder reactors to destroy by transmutation some of the longer-lived non-actinides as well, the waste must be segregated from the environment for one to a few hundred years, and therefore this is properly categorized as a long-term problem. Subcritical reactors or fusion reactors could also reduce the time the waste has to be stored. It has been argued that the best solution for the nuclear waste is above ground temporary storage since technology is rapidly changing. The current waste may well become a valuable resource in the future.

In countries with nuclear power, radioactive wastes comprise less than 1% of total industrial toxic wastes, which remain hazardous indefinitely unless they decompose or are treated so that they are less toxic or, ideally, completely non-toxic.

The nuclear industry also produces a volume of low-level radioactive waste in the form of contaminated items like clothing, hand tools, water purifier resins, and (upon decommissioning) the materials of which the reactor itself is built. In the United States, the Nuclear Regulatory Commission has repeatedly attempted to allow low-level materials to be handled as normal waste: landfilled, recycled into consumer items, etc. Most low-level waste releases very low levels of radioactivity and is only considered radioactive waste because of its history. For example, according to the standards of the NRC, the radiation released by coffee is enough to treat it as low level waste. Overall, nuclear power produces far less waste material than fossil-fuel based power plants. Coal-burning plants are particularly noted for producing large amounts of toxic and mildly radioactive ash due to concentrating naturally occurring metals and radioactive material from the coal.

Nuclear Reprocessing

Reprocessing can recover up to 95% of the remaining uranium and plutonium in spent nuclear fuel, putting it into new mixed oxide fuel. This also produces a reduction in long term radioactivity within the remaining waste, since this is largely short-lived fission products, and reduces its volume by over 90%. Reprocessing of civilian fuel from power reactors is currently done on large scale in Britain, France and (formerly) Russia, will be in China and perhaps India, and is being done on an expanding scale in Japan. Iran has announced its intention to complete the nuclear fuel cycle via reprocessing, a move which has led to criticism from the United States and the International Atomic Energy Agency. Unlike other countries, U.S. policy at one stage forbade recycling of used fuel; although this policy was reversed, spent fuel is all currently treated as waste.

Economy

Opponents of nuclear power argue that any of the environmental benefits are outweighed by safety compromises and by the costs related to construction and operation of nuclear power plants, including costs for spent-fuel disposal and plant retirement. Proponents of nuclear power respond that nuclear energy is the only power source which explicitly factors the estimated costs for waste containment and plant decommissioning into its overall cost, and that the quoted cost of fossil fuel plants is deceptively low for this reason. The cost of some renewables would be increased too if they included necessary back-up due to their intermittent nature.

A UK Royal Academy of Engineering report in 2004 looked at electricity generation costs from new plants in the UK. In particular it aimed to develop "a robust approach to compare directly the costs of intermittent generation with more dependable sources of generation". This meant adding the cost of standby capacity for wind, as well as carbon values up to £30 per tonne CO_2 for coal and gas. Wind power was calculated to be more than twice as expensive as nuclear power. Without a carbon tax, the cost of production through coal, nuclear and gas ranged £0.022-0.026/kWh and coal gasification was £0.032/kWh. When carbon tax was added (up to £0.025) coal came close to onshore wind (including back-up power) at £0.054/kWh-offshore wind is £0.072/kWh.

Nuclear power remained at £0.023/kWh either way, as it produces negligible amounts of CO_2. Nuclear figures included decommissioning costs. In one study, certain gas cogeneration plants were calculated to be three to four times more cost-effective than nuclear power, if all the heat produced was used onsite or in a local heating system. However, the study estimated only 25 year plant lifetimes (60 is now common), 68% capacity factors were assumed (above 90% is now common), other conservatisms were applied, and nuclear power also produces heat which could be used in similar ways (although most nuclear power plants are located in remote areas). The study then found similar costs for nuclear power and most other forms of generation if not including external costs (such as back-up power).

Capital Costs

Generally, a nuclear power plant is significantly more expensive to build than an equivalent coal-fuelled or gas-fuelled plant. Coal is significantly more expensive than nuclear fuel, and natural gas significantly more expensive than coal-thus, capital costs aside, natural gas-generated power is the most expensive. However, servicing the capital costs for a nuclear power dominate the costs of nuclear-generated electricity, contributing about 70% of costs (assuming a 10% discount rate).

The recent liberalisation of the electricity market in many countries has made the economics of nuclear power generation less attractive. Previously a monopolistic provider could guarantee output requirements decades into the future. Private generating companies have to accept shorter output contracts and the risks of future competition, so desire a shorter return on investment period which favours generation plants with lower capital costs.

In many countries, licensing, inspection and certification of nuclear power plants has added delays and construction costs to their construction. In the U.S. many new regulations were put in place after the Three Mile Island partial meltdown. Gas-fired and coal-fired plants do not face such regulations. Because a power plant does not yield profits during construction, longer construction times translate directly into higher interest charges on borrowed construction funds. However, the regulatory processes for siting, licensing, and constructing have been standardized since their introduction, to make construction of newer and safer designs more attractive to companies.

In Japan and France, construction costs and delays are significantly diminished because of streamlined government licensing and certification procedures. In France, one model of reactor was type-certified, using a safety engineering process similar to the process used to certify aircraft models for safety. That is, rather than licensing individual reactors, the regulatory agency certified a particular design and its construction process to produce safe reactors. U.S. law permits type-licensing of reactors, a process which is about to be used.

To encourage development of nuclear power, under the Nuclear Power 2010 Program the U.S. Department of Energy (DOE) has offered interested parties the opportunity to introduce France's model for licensing and to subsidize 25% to 50% of the construction cost overruns due to delays for the first six new plants. Several applications were made, two sites have been chosen to receive new plants, and other projects are pending.

Operating Costs

In general, coal and nuclear plants have the same types of operating costs (operations and maintenance plus fuel costs). However, nuclear and coal differ in the relative size of those costs. Nuclear has lower fuel costs but higher operating and maintenance costs. In recent times in the United States savings due to lower fuel cost have not been low enough for nuclear to repay its higher investment cost.

Thus no new nuclear reactors have been ordered in the United States since the 1970s. Coal's operating cost advantages have only rarely been sufficient to encourage the construction of new coal based power generation. Around 90 to 95 percent of new power plant construction in the United States has been natural gas-fired.

To be competitive in the current market, both the nuclear and coal industries must reduce new plant investment costs and construction time. The burden is clearly greater for nuclear producers than for coal producers, because investment costs are higher for nuclear plants. Operation and maintenance costs are particularly important because they represent a large portion of costs for nuclear power. One of the primary reasons for the uncompetitiveness of the U.S. nuclear industry has been the lack of any measure that provides an economic incentive to reduce carbon emissions (carbon tax). Many economists and environmentalists have called for a mechanism to take into account the negative externalities of coal and gas consumption. In such an environment, it is argued that nuclear will become cost-competitive in the United States.

Subsidies

Critics of nuclear power claim that it is the beneficiary of inappropriately large economic subsidies-mainly taking the forms of taxpayer-funded research and development and limitations on disaster liability-and that these subsidies, being subtle and indirect, are often overlooked when comparing the economics of nuclear against other forms of power generation. However, competing energy sources also receive subsidies. Fossil fuels receive large direct and indirect subsidies, such as tax benefits and not having to pay for the Greenhouse gases they emit. Renewables receive large direct production subsidies and tax breaks in many nations.

Energy research and development (R&D) for nuclear power has and continues to receive much larger state subsidies than R&D for renewable energy or fossil fuels. However, today most of this takes places in Japan and France: in most other nations renewable R&D get more money. In the U.S., public research money for nuclear fission declined from 2,179 to 35 million dollars between 1980 and 2000. However, in order to restart the industry, the next six U.S. reactors will receive subsidies equal to those of renewables and, in the event of cost overruns due to delays, at least partial compensation for the overruns.

According to the DOE, insurance for nuclear or radiological incidents in the U.S., is subsidized by the Price-Anderson Nuclear Industries Indemnity Act. In July 2005, Congress extended this Act to newer facilities. In the UK, the Nuclear Installations Act of 1965 governs liability for nuclear damage for which a UK nuclear licensee is responsible. The Vienna Convention on Civil Liability for Nuclear Damage puts in place an international framework for nuclear liability.

Other Economic Issues

Nuclear Power plants tend to be most competitive in areas where other fuel resources are not readily available-France, most notably, has almost no native supplies of fossil fuels. The province of Ontario, Canada is already using all of its best sites for hydroelectric power, and has minimal supplies of fossil fuels, so a number of nuclear plants have been built there. India is also building new nuclear plants to supplement its vast coal reserves and coal-generated electricity. Conversely, in the United Kingdom, according to the government's Department Of Trade And Industry, no further nuclear power stations are to be built, due to the high cost per unit of nuclear power compared to fossil fuels. However, the British government's chief scientific advisor David King reports that building one more generation of nuclear power plants may be necessary. China tops the list of planned new plants, due to its rapidly expanding economy and fervent construction in many types of energy projects.

Most new gas-fired plants are intended for peak supply. The larger nuclear and coal plants cannot quickly adjust their instantaneous power production, and are generally intended for baseline supply. The market price for baseline power has not increased as rapidly as that for peak demand. Some new experimental reactors, notably pebble bed modular reactors, are specifically designed for peaking power.

Any effort to construct a new nuclear facility around the world, whether an older design or a newer experimental design, must deal with NIMBY objections. Given the high profile of both the Three Mile Island and Chernobyl accidents, few municipalities welcome a new nuclear reactor, processing plant, transportation route, or experimental nuclear burial ground within their borders, and many have issued local ordinances prohibiting the development of nuclear power. However, a few U.S. areas with nuclear units are campaigning for more.

Current nuclear reactors return around 40-60 times the invested energy when using life cycle analysis. This is better than coal, natural gas, and current renewables except hydropower.

The Rocky Mountain Institute gives other reasons why nuclear power plants may not be economical. In the U.S. this includes long lead times on risky investments, and the more cost-effective approach of investing in efficiency instead of new power plants. Nuclear power, coal, and wind power are currently the only realistic large scale energy sources that would be able to replace oil and natural gas after a peak in global oil and gas production has been reached. However, The Rocky Mountain Institute claims that in the U.S. increases in transportation efficiency and stronger, lighter cars would replace most oil usage with what it calls negawatts. Biofuels can then substitute for a significant portion of the remaining oil use. Efficiency, insulation, solar thermal, and solar photovoltaic technologies can substitute for most natural gas usage after a peak in production.

Nuclear proponents often assert that renewable sources of power have not solved problems like intermittent output, high costs, and diffuse output which requires the use of large surface areas and much construction material and which increases distribution losses. For example, studies in Britain have shown that increasing wind power production contribution to 20% of all energy production, without costly pumped hydro or electrolysis/fuel cell storage, would only reduce coal or nuclear power plant capacity by 6.7% (from 59 to 55 GWe) since they must remain as backup in the absence of power storage. Nuclear proponents often claim that increasing the contribution of intermittent energy sources above that is not possible with current technology.

Some renewable energy sources, such as solar, overlap well with peak electrical production and reduce the need of spare generating capacity. Future applications that use electricity when it is available (e.g. for pressurizing water systems, desalination, or hydrogen generation) would help to reduce the spare generation capacity required by both nuclear and renewable energy sources.

COMPONENTS OF ENVIRONMENTAL SCIENCE

Atmospheric sciences focuses on the Earth's atmosphere, with an emphasis upon its interrelation to other systems. Atmospheric sciences can include studies of meteorology, greenhouse gas phenomena, atmospheric dispersion modeling of airborne contaminants, sound propagation phenomena related to noise pollution, and even light pollution.

Taking the example of the global warming phenomena, physicists create computer models of atmospheric circulation and infra-red radiation transmission, chemists examine the inventory of atmospheric chemicals and their reactions, biologists analyse the plant and animal contributions to carbon dioxide fluxes, and specialists such as meteorologists and oceanographers add additional breadth in understanding the atmospheric dynamics.

Ecology. An interdisciplinary analysis of an ecological system which is being impacted by one or more stressors might include several related environmental science fields. For example, one might examine an estuarine setting where a proposed industrial development could impact certain species by water and air pollution. For this study, biologists would describe the flora and fauna, chemists would analyse the transport of water pollutants to the marsh, physicists would calculate air pollution emissions and geologists would assist in understanding the marsh soils and bay muds.

Environmental chemistry is the study of chemical alterations in the environment. Principal areas of study include soil contamination and water pollution. The topics of analysis include chemical degradation in the environment, multi-phase transport of chemicals (for example, evaporation of a solvent containing lake to yield solvent as an air pollutant), and chemical effects upon biota. As an example study, consider the case of a leaking solvent tank which has entered the habitat soil of an endangered species of amphibian. As a method to resolve or understand the extent of soil contamination and subsurface transport of solvent, a computer model would be implemented.

Chemists would then characterize the molecular bonding of the solvent to the specific soil type, and biologists would study the impacts upon soil arthropods, plants, and ultimately pond-dwelling organisms that are the food of the endangered amphibian.

Geosciences include environmental geology, enviro-nmental soil science, volcanic phenomena and evolution of the Earth's crust. In some classification systems this can also include hydrology, including oceanography.

As an example study of soils erosion, calculations would be made of surface run-off by soil scientists. Fluvial geomorphologists would assist in examining sediment transport in overland flow. Physicists would contribute by assessing the changes in light transmission in the receiving waters. Biologists would analyse subsequent impacts to aquatic flora and fauna from increases in water turbidity.

Environmental chemistry is the scientific study of the chemical and biochemical phenomena that occur in natural places. It should not be confused with green chemistry, which seeks to reduce potential pollution at its source. It can be defined as the study of the sources, reactions, transport, effects, and fates of chemical species in the air, soil, and water environments; and the effect of human activity on these. Environmental chemistry is an interdisciplinary science that includes atmospheric, aquatic and soil chemistry, as well as heavily relying on analytical chemistry and being related to environmental and other areas of science.

Environmental chemistry involves first understanding how the uncontaminated environment works, which chemicals in what concentrations are present naturally, and with what effects. Without this it would be impossible to accurately study the effects humans have on the environment through the release of chemicals. Environmental chemists draw on a range of concepts from chemistry and various environmental sciences to assist in their study of what is happening to a chemical species in the environment. Important general concepts from chemistry include understanding chemical reactions and equations, solutions, units, sampling, and analytical techniques.

DISTRIBUTION OF TEMPERATURE

The pull, as we call it, of gravity makes water run down hill. It also makes a heavy liquid underrun a lighter one in the same level; both are drawn in the direction of the bottom, but the pull on the heavier, or denser, is greater than on the lighter, and the stronger pull prevails. Gravity also makes an isolated mass of liquid or gas in a heavier one go up, not down; it is pushed or buoyed up by a force equal to the difference between the weight of the lighter and that of an equal volume of the heavier.

Clearly, then, whenever two masses of air of unequal density come into free contact with each other the lighter is pushed up and away, except in the case of properly adjusted winds. Now air rapidly increases in volume, and correspondingly decreases in density, with increase of temperature—roughly

1% per 5° F. at ordinary temperatures. Hence the hot air in a chimney is lighter than an equal volume of the cold air on the outside, and therefore is pushed up by the latter which, in turn, is heated and itself pushed up, and so on as long as there is a fire in the grate to supply the heat.

To be sure, the combustion alters the composition of the air (makes it richer in carbon dioxide if coal is used, and in both carbon dioxide and water vapour if wood or gas is the fuel, and poorer in oxygen) in such manner as to render that in the chimney heavier, at the same temperature, than that outside, but this increase in density through change in composition is small in comparison to its decrease in density by heating.

At most it could balance or offset a temperature increase of only about 40° F. over coal, or 10° F. over wood, while ordinarily the effect if much less, since commonly only part of the oxygen is consumed; hence the heating, being several times this maximum value, has, in any case, the best of the argument, as it were, and the chimneys keep on drawing. Similarly, air in the open is underrun and pushed up by even slightly cooler adjacent air of the same composition, unless, as already explained, the two masses happen to be flowing past each other in the right positions and directions and with the proper velocities.

Actually, the heated air expands as its temperature rises, and overflows above wherever its pressure is thus made greater than that of the adjacent atmosphere. This overflow, or outflow, decreases the pressure at the bottom, and in the lower portions, of the heated air, and at the same time increases the pressure round about under the places of overflow—mass, hence weight, is removed from one place and added to others.

This disturbs the balance. Gravity tends to restore it and thereby induces winds in the direction, initially at least, of higher to lower pressure. If the heated region is very small, equilibrium is quickly established, unless the heating is maintained. But where the higher temperature covers a large the winds no longer flow directly towards the centre of lowest pressure but more or less round about it, owing to the rotation of the earth, in a manner seemingly most contrarious.

This heating in innumerable cases is very local and of only a few hours' duration; in many others it is quite extensive and lasts days Weeks, and even all season long; while its greatest manifestation is the year after year and age after age continuously higher temperature in tropical realms and lower in the frigid zones. This perpetual heating of the atmosphere over one great region, and its ceaseless cooling over another, or rather, two others, keeps it continuously out of balance and makes the winds, especially the trades and the westerlies, forever to blow—to blow dizzily over a rotating earth, and time and again violently and confusedly incident to the rapid, the all but explosive, delivery, by condensation, to a limited region of vast quantities of heat that had been slowly accumulated by evaporation from others afar off.

The whole of the atmosphere to the tops of the highest clouds, that is, the whole of the troposphere, is a huge convection system, greatly complicated by the rotation of the earth and all but hopelessly confused by evaporation and condensation. The stratosphere, too, has its circulation, but as yet not much is known about it.

Of course it is difference in pressure *at the same level* that pushes the air about, or makes the winds to blow, but, as explained, this difference in pressure depends, in turn, mainly (water vapour has a little to do with it) on the distribution of temperature. That is one reason, but not the only one, why this distribution is so important. Perhaps some good physicist will insist that it really isn't difference of pressure at the same height above sea level that makes the winds blow, but difference of pressure over an "isentropic surface," or surface of "equal entropy."

Well, he would be right in respect to appreciable heights above the surface, because for the free air the isentropic surface is the "level" surface. But nothing short of a surgical operation can get the idea of entropy into the other fellow's head, and there is no rivet, weld, or hermetic seal that will keep it there. Besides, commonly (not always), there isn't much difference between the two after all—"same level" and "isentropic level"—and so we will stick to the one everybody knows and no one forgets, that is, "same level."

Source of Heat

When we think of the source of heat, especially in the winter-time, we are likely to have in mind some sort of combustion, for that is the cause of the tropical climate we have indoors at that season. But indoors is a mighty small place in comparison with all outdoors; and outdoors is heated, too, often very hot in summer, and always far above the 460° below zero Fahrenheit that would be its temperature if there were no heating at all. Almost every bit of this enormous amount of heating comes from just one source, the sun.

Incoming Radiation

The radiation from the sun is so great that if it all got through the atmosphere enough would fall on each square foot directly facing it to heat a gallon of water from the freezing point to the boiling point in just three and a half hours. But it does not all get through, and what does get through always comes in slopingly except wherever the sun happens for the moment to be directly overhead.

In fact, owing to the reflecting power of clouds, especially, and the surface of the earth, and to the scattering (not reflection) of light by the molecules of the air and by the myriads of dust motes, one-third, roughly, of the incoming radiation is thrown off to space without producing any effect whatever on the temperature of the atmosphere or of the earth beneath. Another one-third, again roughly, of the incoming solar radiation is absorbed by the atmosphere, and the remaining portion by the earth.

These statements apply to the earth as a whole. The ratios between loss by reflection and scattering, air absorption, and earth absorption, vary widely from place to place and season to season, owing mainly to differences in humidity, cloudiness, and elevation of the sun above the horizon, and differences in the character of the surface of the earth—whether land, water, snow or ice, bare soil or vegetation.

Clear Sky Radiations

It is interesting to note that the amount of radiation reaching the earth from a clear sky is equal to a considerable fraction of that which reaches it from the sun directly. At sea level the amount of sky radiation onto a horizontal surface of any particular size, a square foot, say, is equal to about 7.8% of the amount of unaffected, or direct, solar radiation onto an equal area squarely facing the sun at the same time and place. When the sun is directly overhead its supply of heat to a horizontal surface at sea level is nearly 13 times as great as that from the sky. When it is one-third of the way down from the zenith to the horizon its contribution of heat to the earth is only a little more than 6 times that from the sky, and each is then decidedly less than it is when the sun is in the zenith. Finally, the two sources are equal, though both are still further enfeebled, when the sun is above the horizon about one-twelfth the distance to the zenith.

The brightness of the clear sky is greatest near the sun, as even casual observations readily show, and decreases gradually with increase of distance therefrom over a large part of the whole area. Hence the total of sky radiation received per minute on a horizontal surface is greatest at noon, as is also the direct solar radiation. The intensity of sky radiation decreases, in general, with increase of height above sea level, while that of the direct solar radiation increases.

6

Global Environmental Change and Biological Controls

Today the activities of one species, humans, are reducing the diversity of all others and transforming the global environment. Ecosystems subjected to the stresses of "global change" (including climate change and altered weather patterns, the depletion of stratospheric ozone, deforestation, coastal pollution, and marked reductions of biological diversity) become more susceptible to the emergence, invasion, and spread of opportunistic species.

When subject to multiple stresses, natural environments can exhibit symptoms that indicate reductions in resilience, resistance, and regenerative capabilities. Conversely, ecosystems have inherent flexibilities and survival strategies that can be strengthened by systematic stress, such as the seasonal battering they must endure in temperate latitudes. But their tolerance for abuse has its limits. Several features of global change tend to reduce predators disproportionately, and in the process release prey from their biological controls. Among the most widespread are:

- Fragmentation and loss of habitat
- Dominance of monocultures in agriculture and aquaculture
- Excessive use of toxic chemicals
- Increased ultraviolet radiation, and
- Climate change and weather instability.

The breaking up of large tracts of forest or other natural wilderness into smaller and more diverse patches reduces the available habitat for large predators, and favours many pests. Land and climate changes may act synergistically, as when constricted habitat frustrates a species' ability to migrate north or south to survive altered climatic conditions. Extensive deforestation and climate anomalies—such as the delayed monsoon rains that resulted from this year's El Nino—can also act synergistically, with costly results. A ready example is the massive haze from burning that covered much of Southeast Asia in September and October, causing acute and chronic respiratory damage and losses in trade, investment, and tourism—the latter, a $26 billion a year industry.

The dedication of land to *monoculture,* that is, the cultivation of single crops with restricted genetic and species diversity, renders plants more vulnerable to disease. Simplified systems are also more susceptible to climatic extremes and to outbreaks of pests.

Over-use of pesticides kills birds and beneficial insects, as noted in 1962 by Rachel Carson. The title of her book, Silent Spring, made reference to the absence of the chorus of birds in springtime, and the resulting resurgence of plant-eating insects—that had also evolved a resistance to pesticides. The worldwide response to her message transformed agricultural policies and generated more enlightened pest management. But today, the heavy application of pesticides still carries risks to both human health and natural systems. Over-use of pesticides in Texas and Alabama to control the boll weevil has alarmed farmers, for friendly insects such as spiders and lady bugs have died off and other plant pests have rebounded.

Ecosystem Health

As noted earlier, one of "nature's services" is to keep opportunistic species under control. Maintaining this service entails sustaining the health and integrity of ecosystems. One of the essentials is genetic and species biodiversity to provide alternative hosts for disease organisms. Another is sufficient stability among functional groups of species (such as recyclers, scavengers, predators, competitors, and prey) to ensure the suppression of opportunists and preserve essential ecological functions. Habitat is crucial.

Stands of trees interspersed with agricultural fields, for instance, support birds that control insects; clean ponds with healthy populations of fish serve to control mosquito larvae; and adequate wetlands filter excess nutrients, harmful chemicals, and microorganisms.

As a case in point, in tidewater Maryland, buffer zones around farms and the restoration of wetlands and river-bed trees can absorb the flow of sediments, chemicals, and harmful organisms into Chesapeake Bay, and thus reduce the emergence and spread of algae, toxic to fish. Ecosystems are also interrelated: healthy forests and mangroves in Central America, for example, are crucial to coral reefs that spawn fish stocks, formed at the origin of the great Gulf Stream.

Maintaining the integrity of natural environmental systems provides generalized defenses against the proliferation of opportunistic pests and disease.

Population explosions of nuisance organisms, be they animals or plants or microbes, often reflect failing ecosystem health: a sign of systems out of equilibrium, in terms of the balance of organisms required to perform essential functions. The damage done, moreover, can be cumulative, for multiply-stressed systems are less able to resist and rebound when other stresses come along.

Rodents, insects, and algae are thus key biological indicators of ecosystem health. Their populations and species compositions respond rapidly to environmental change—particularly to an increase in their food supply, or a drop in the number of their natural predators. These indicator species are also linked to human health.

Impacts of a Loss of Biodiversity

The present rate of species extinctions around the world is a potential threat to human health when one considers the role that predators play in containing infectious disease. From the largest to the smallest scales, an essential element in natural systems for countering stress is a diversity of defenses and responses. Thus, animals that seem redundant may serve as "insurance" species in a natural ecosystem, providing a back-up layer of resilience and resistance when others are lost from disease, a changing environment, or a shortage of food or water.

In 1996 the World Conservation Union reported that one-fourth of all species of mammals—and similar proportions of reptiles, amphibians, and fish—are threatened. The current rate of extinctions (estimated at 100 to 1,000 times the rate of loss in the pre-human era) falls heaviest on large predators and "specialists," and thus may initially favour the spread of opportunistic species.

ECOLOGICAL SYSTEMS

In terrestrial ecosystems, the earlier timing of spring events, and poleward and upward shifts in plant and animal ranges, have been linked with high confidence to recent warming. Future climate change is expected to particularly affect certain ecosystems, including tundra, mangroves, and coral reefs. It is expected that most ecosystems will be affected by higher atmospheric CO_2 levels, combined with higher global temperatures. Overall, it is expected that climate change will result in the extinction of many species and reduced diversity of ecosystems.

Biodiversity

Deforestation on a human scale results in declines in biodiversity and on a natural global scale is known to cause the extinction of many species. The removal or destruction of areas of forest cover has resulted in a degraded environment with reduced biodiversity. Forests support biodiversity, providing habitat for wildlife; moreover, forests foster medicinal conservation. With forest biotopes being irreplaceable source of new drugs (such as taxol), deforestation can destroy genetic variations (such as crop resistance) irretrievably. Since the tropical rainforests are the most diverse ecosystems on Earth and about 80 per cent of the world's known biodiversity could be found in tropical rainforests, removal or destruction of significant areas of forest cover has resulted in a degraded environment with reduced biodiversity.

It has been estimated that we are losing 137 plant, animal and insect species every single day due to rainforest deforestation, which equates to 50,000 species a year. Others state that tropical rainforest deforestation is contributing to the ongoing Holocene mass extinction. The known extinction rates from deforestation rates are very low, approximately 1 species per year from mammals and birds which extrapolates to approximately 23,000 species per year for all species. Predictions have been made that more than 40 per cent of the animal and plant species in Southeast Asia could be wiped out in the 21st century. Such predictions were called into question by 1995 data that show that within regions of Southeast Asia much of the original forest has been converted to monospecific plantations, but that potentially endangered species are few and tree flora remains widespread and stable.

Scientific understanding of the process of extinction is insufficient to accurately make predictions about the impact of deforestation on biodiversity. Most predictions of forestry related biodiversity loss are based on species-area models, with an underlying assumption that as the forest declines species diversity will decline similarly. However, many such models have been proven to be wrong and loss of habitat does not necessarily lead to large scale loss of species. Species-area models are known to overpredict the number of species known to be threatened in areas where actual deforestation is ongoing, and greatly overpredict the number of threatened species that are widespread.

Hydrological

The water cycle is also affected by deforestation. Trees extract groundwater through their roots and release it into the atmosphere. When part of a forest is removed, the trees no longer evaporate away this water, resulting in a much drier climate. Deforestation reduces the content of water in the soil and groundwater as well as atmospheric moisture. The dry soil leads to lower water intake for the trees to extract. Deforestation reduces soil cohesion, so that erosion, flooding and landslides ensue. Shrinking forest cover lessens the landscape's capacity to intercept, retain and transpire precipitation. Instead of trapping precipitation, which then percolates to groundwater systems, deforested areas become sources of surface water runoff, which moves much faster than subsurface flows. That quicker transport of surface water can translate into flash flooding and more localised floods than would occur with the forest cover. Deforestation also contributes to decreased evapotranspiration, which lessens atmospheric moisture which in some cases affects precipitation levels downwind from the deforested area, as water is not recycled to downwind forests, but is lost in runoff and returns directly to the oceans. According to one study, in deforested north and northwest China, the average annual precipitation decreased by one third between the 1950s and the 1980s.

Trees, and plants in general, affect the water cycle significantly:

- their canopies intercept a proportion of precipitation, which is then evaporated back to the atmosphere (canopy interception);
- their litter, stems and trunks slow down surface runoff;
- their roots create macropores – large conduits – in the soil that increase infiltration of water;
- they contribute to terrestrial evaporation and reduce soil moisture via transpiration;
- their litter and other organic residue change soil properties that affect the capacity of soil to store water.
- their leaves control the humidity of the atmosphere by transpiring. 99 per cent of the water absorbed by the roots moves up to the leaves and is transpired.

As a result, the presence or absence of trees can change the quantity of water on the surface, in the soil or groundwater, or in the atmosphere. This in turn changes erosion rates and the availability of water for either ecosystem functions or human services. The forest may have little impact on flooding in the case of large rainfall events, which overwhelm the storage capacity of forest soil if the soils are at or close to saturation. Tropical rainforests produce about 30 per cent of our planet's fresh water.

Soil

Undisturbed forests have a very low rate of soil loss, approximately 2 metric tons per square kilometre (6 short tons per square mile). Deforestation generally increases rates of soil erosion, by increasing the amount of runoff and reducing the protection of the soil from tree litter.

This can be an advantage in excessively leached tropical rain forest soils. Forestry operations themselves also increase erosion through the development of roads and the use of mechanised equipment.

China's Loess Plateau was cleared of forest millennia ago. Since then it has been eroding, creating dramatic incised valleys, and providing the sediment that gives the Yellow River its yellow colour and that causes the flooding of the river in the lower reaches (hence the river's nickname 'China's sorrow').

Removal of trees does not always increase erosion rates. In certain regions of southwest US, shrubs and trees have been encroaching on grassland. The trees themselves enhance the loss of grass between tree canopies. The bare intercanopy areas become highly erodible. The US Forest Service, in Bandelier National Monument for example, is studying how to restore the former ecosystem, and reduce erosion, by removing the trees. Tree roots bind soil together, and if the soil is sufficiently shallow they act to keep the soil in place by also binding with underlying bedrock. Tree removal on steep slopes with shallow soil thus increases the risk of landslides, which can threaten people living nearby. However most deforestation only affects the trunks of trees, allowing for the roots to stay rooted, negating the landslide.

Forest Transition Theory

The forest area change may follow a pattern suggested by the forest transition (FT) theory, whereby at early stages in its development a country is characterised by high forest cover and low deforestation rates (HFLD countries).

Then deforestation rates accelerate (HFHD, high forest cover – high deforestation rate), and forest cover is reduced (LFHD. low forest cover – high deforestation rate), before the deforestation rate slows (LFLD, low forest cover – low deforestation rate), after which forest cover stabilises and eventually starts recovering. FT is not a "law of nature," and the pattern is influenced by national context (for example, human population density, stage of development, structure of the economy), global economic forces, and government policies. A country may reach very low levels of forest cover before it stabilises, or it might through good policies be able to "bridge" the forest transition.

FT depicts a broad trend, and an extrapolation of historical rates therefore tends to underestimate future BAU deforestation for counties at the early stages in the transition (HFLD), while it tends to overestimate BAU deforestation for countries at the later stages (LFHD and LFLD).

Countries with high forest cover can be expected to be at early stages of the FT. GDP per capita captures the stage in a country's economic development, which is linked to the pattern of natural resource use, including forests. The choice of forest cover and GDP per capita also fits well with the two key scenarios in the FT:

(i) a forest scarcity path, where forest scarcity triggers forces (for example, higher prices of forest products) that lead to forest cover stabilisation; and

(ii) an economic development path, where new and better off-farm employment opportunities associated with economic growth (= increasing GDP per capita) reduce profitability of frontier agriculture and slows deforestation.

Historical Causes

Prehistory: The Carboniferous Rainforest Collapse, was an event that occurred 300 million years ago. Climate change devastated tropical rainforests causing the extinction of many plant and animal species. The change was abrupt, specifically, at this time climate became cooler and drier, conditions that are not favourable to the growth of rainforests and much of the biodiversity within them. Rainforests were fragmented forming shrinking 'islands' further and further apart. This sudden collapse affected several large groups, effects on amphibians were particularly devastating, while reptiles fared better, being ecologically adapted to the drier conditions that followed. Rainforests once covered 14 per cent of the earth's land surface; now they cover a mere 6 per cent and experts estimate that the last remaining rainforests could be consumed in less than 40 years. Small scale deforestation was practiced by some societies for tens of thousands of years before the beginnings of civilization. The first evidence of deforestation appears in the Mesolithic period. It was probably used to convert closed forests into more open ecosystems favourable to game animals. With the advent of agriculture, larger areas began to be deforested, and fire became the prime tool to clear land for crops. In Europe there is little solid evidence before 7000 BC. Mesolithic foragers used fire to create openings for red deer and wild boar. In Great Britain, shade-tolerant species such as oak and ash are replaced in the pollen record by hazels, brambles, grasses and nettles. Removal of the forests led to decreased transpiration, resulting in the formation of upland peat bogs. Widespread decrease in elm pollen across Europe between 8400–8300 BC and 7200–7000 BC, starting in southern Europe and gradually moving north to Great Britain, may represent land clearing by fire at the onset of Neolithic agriculture.

The Neolithic period saw extensive deforestation for farming land. Stone axes were being made from about 3000 BC not just from flint, but from a wide variety of hard rocks from across Britain and North America as well. They include the noted Langdale axe industry in the English Lake District, quarries developed at Penmaenmawr in North Wales and numerous other locations. Rough-outs were made locally near the quarries, and some were polished locally to give a fine finish. This step not only increased the mechanical strength of the axe, but also made penetration of wood easier. Flint was still used from sources such as Grimes Graves but from many other mines across Europe. Evidence of deforestation has been found in Minoan Crete; for example the environs of the Palace of Knossos were severely deforested in the Bronze Age.

Pre-industrial History

Throughout most of history, humans were hunter gatherers who hunted within forests. In most areas, such as the Amazon, the tropics, Central America, and the Caribbean, only after shortages of wood and other forest products occur are policies implemented to ensure forest resources are used in a sustainable manner.

In ancient Greece, Tjeered van Andel and co-writers summarised three regional studies of historic erosion and alluviation and found that, wherever adequate evidence exists, a major phase of erosion follows, by about 500-1,000 years the introduction of farming in the various regions of Greece, ranging from the later Neolithic to the Early Bronze Age. The thousand years following the mid-first millennium BCE saw serious, intermittent pulses of soil erosion in numerous places. The historic silting of ports along the southern coasts of Asia Minor (*e.g.* Clarus, and the examples of Ephesus, Priene and Miletus, where harbors had to be abandoned because of the silt deposited by the Meander) and in coastal Syria during the last centuries BC.

Easter Island has suffered from heavy soil erosion in recent centuries, aggravated by agriculture and deforestation. Jared Diamond gives an extensive look into the collapse of the ancient Easter Islanders in his book *Collapse*. The disappearance of the island's trees seems to coincide with a decline of its civilization around the 17th and 18th century. He attributed the collapse to deforestation and over-exploitation of all resources. The famous silting up of the harbor for Bruges, which moved port commerce to Antwerp, also followed a period of increased settlement growth (and apparently of deforestation) in the upper river basins.

In early medieval Riez in upper Provence, alluvial silt from two small rivers raised the riverbeds and widened the floodplain, which slowly buried the Roman settlement in alluvium and gradually moved new construction to higher ground; concurrently the headwater valleys above Riez were being opened to pasturage. A typical progress trap was that cities were often built in a forested area, which would provide wood for some industry (for example, construction, shipbuilding, pottery). When deforestation occurs without proper replanting, however; local wood supplies become difficult to obtain near enough to remain competitive, leading to the city's abandonment, as happened repeatedly in Ancient Asia Minor. Because of fuel needs, mining and metallurgy often led to deforestation and city abandonment.

With most of the population remaining active in (or indirectly dependent on) the agricultural sector, the main pressure in most areas remained land clearing for crop and cattle farming. Enough wild green was usually left standing (and partially used, for example, to collect firewood, timber and fruits, or to graze pigs) for wildlife to remain viable. The elite's (nobility and higher clergy) protection of their own hunting privileges and game often protected significant woodlands.

Major parts in the spread (and thus more durable growth) of the population were played by monastical 'pioneering' (especially by the Benedictine and Commercial orders) and some feudal lords' recruiting farmers to settle (and become tax payers) by offering relatively good legal and fiscal conditions. Even when speculators sought to encourage towns, settlers needed an agricultural belt around or sometimes within defensive walls. When populations were quickly decreased by causes such as the Black Death or devastating warfare (for example, Genghis Khan's Mongol hordes in eastern and central Europe, Thirty Years' War in Germany), this could lead to settlements being abandoned. The land was reclaimed by nature, but the secondary forests usually lacked the original biodiversity.

From 1100 to 1500 AD, significant deforestation took place in Western Europe as a result of the expanding human population. The large-scale building of wooden sailing ships by European (coastal) naval owners since the 15th century for exploration, colonisation, slave trade–and other trade on the high seas consumed many forest resources. Piracy also contributed to the over harvesting of forests, as in Spain. This led to a weakening of the domestic economy after Columbus' discovery of America, as the economy became dependent on colonial activities (plundering, mining, cattle, plantations, trade, etc.)

In *Changes in the Land* (1983), William Cronon analysed and documented 17th-century English colonists' reports of increased seasonal flooding in New England during the period when new settlers initially cleared the forests for agriculture. They believed flooding was linked to widespread forest clearing upstream.

The massive use of charcoal on an industrial scale in Early Modern Europe was a new type of consumption of western forests; even in Stuart England, the relatively primitive production of charcoal has already reached an impressive level. Stuart England was so widely deforested that it depended on the Baltic trade for ship timbers, and looked to the untapped forests of New England to supply the need. Each of Nelson's Royal Navy war ships at Trafalgar (1805) required 6,000 mature oaks for its construction. In France, Colbert planted oak forests to supply the French navy in the future. When the oak plantations matured in the mid-19th century, the masts were no longer required because shipping had changed.

Norman F. Cantor's summary of the effects of late medieval deforestation applies equally well to Early Modern Europe:

> *Europeans had lived in the midst of vast forests throughout the earlier medieval centuries. After 1250 they became so skilled at deforestation that by 1500 they were running short of wood for heating and cooking. They were faced with a nutritional decline because of the elimination of the generous supply of wild game that had inhabited the now-disappearing forests, which throughout medieval times had provided the staple of their carnivorous high-protein diet. By 1500 Europe was on the edge of a fuel and nutritional disaster [from] which it was saved in the sixteenth century only by the burning of soft coal and the cultivation of potatoes and maize.*

MARINE COASTAL ECOSYSTEMS

Seashores throughout the world are subject to increasing pressures from residential, recreational, and commercial development. These stresses may become more severe, for human population in the vicinity of sea-coasts is growing at twice the inland rate. Some of the pressures that we exert on coastal ecosystems are summarized in the accompanying box. All can increase the growth of algae. Among the possible consequences of disruption in almost any marine ecosystem is an increase in the opportunistic pathogens that can abet the spread of human disease, sometimes to widespread proportions. One example is cholera.

Cholera

We often think of our modern world as cleansed of the epidemic scourges of ages past. But cholera —an acute and sometimes fatal disease that is accompanied by severe diarrhea— affects more nations today than ever before. The Seventh Pandemic began when the El Tor strain left its traditional home in the Bay of Bengal in the 1960s, travelled to the east and west across Asia, and in the 1970s penetrated the continent of Africa. In 1991, the cholera pandemic reached the Americas, and during the first eighteen months more than half a million cases were reported in Latin America, with 5,000 deaths. Rapid institution of oral rehydration treatment—with clean water, sugar, and salts—limited fatalities in the Americas to about one in a hundred cases. The epidemics also had serious economic consequences. In 1991, Peru lost $770 million in seafood exports and another $250 million in lost tourist revenues because of the disease.

The microbe that transmits cholera, Vibrio cholerae, is found in a dormant or "hibernating" state in algae and microscopic animal plankton, where it can be identified using modern microbiological techniques. But once introduced to people—by drinking contaminated water or eating contaminated fish or shellfish— cholera can recycle through a population, when sewage is allowed to mix with the clean water supply.

Five years ago, in late 1992, a new strain of Vibrio cholerae—O139 Bengal—emerged in India along the coast of the Bay of Bengal. With populations unprotected by prior immunities, this hardy strain quickly spread through adjoining nations, threatening to become the agent of the world's Eighth Cholera Pandemic. For a time, in 1994, El Tor regained dominance. But by 1996, O139 Bengal had reasserted itself. The emergence of this new disease, like all others, involved the interplay of microbial, human host, and environmental factors.

The largest and most intense outbreak of cholera ever recorded occurred in Rwanda in 1994, killing over 40,000 people in the space of weeks, in a nation already ravaged by civil war and ethnic strife. The tragedy of cholera in Rwanda is a reminder of the impacts of conflict and political instability on public health and biological security—just as epidemics may, in turn, contribute to political and economic stability.

Is The Ocean Warming?

Surface temperatures of the ocean have warmed this century, and a gradual warming of the deep ocean has been found in recent years in oceanographic surveys carried out in the tropical Pacific, Atlantic, and Indian Oceans, and at both poles of the Earth. These findings could be indicative of a long-term trend. Corresponding temperature measurements of the sub-surface earth, in cores drilled deep into the Arctic tundra, show a similar effect.

The water that evaporates from warmer seas, and from vegetation and soils of a warmer land surface, intensifies the rate at which water cycles from ocean to clouds and back again. In so doing it increases humidity and reinforces the greenhouse effect. Warm seas are the engines that drive tropical storms and fuel the intensity of hurricanes. More high clouds can also contribute to warmer nights by trapping out-going radiation.

Some Biological Impacts

A warmer ocean can also harm marine plankton, and thus affect more advanced forms of life in the sea. A northward shift in marine flora and fauna along the California coast that has been underway since the 1930s has been associated with the long-term warming of the ocean over that span of time.

Warming—when sufficient nutrients are present—may also be contributing to the proliferation of coastal algal blooms. Harmful algal blooms of increasing extent, duration, and intensity—and involving novel, toxic species—have been reported around the world since the 1970s. Indeed, some scientists feel that the worldwide increase in coastal algal blooms may be one of the first biological signs of global environmental change.

Warm years may result in a confluence of adverse events. The 1987 El Nino was associated with the spread and new growth of tropical and temperate species of algae in higher northern and southern latitudes. Many were toxic algal blooms. In 1987, following a shoreward intrusion of Gulf Stream eddies, the dinoflagellate Gymnodimuim breve, previously found only as far north as the Gulf of Mexico, bloomed about 700 miles north, off Cape Hatteras, North Carolina, where it has since persisted, albeit at low levels. Forty-eight cases of neurological shellfish poisoning occurred in 1987, resulting in an estimated $25 million loss to the seafood industry and the local community. In the same year, anomalous rain patterns and warm Gulf Stream eddies swept unusually close to Prince Edward Island in the Gulf of St. Lawrence. The result, combined with the run-off of local pollutants after heavy rains, was a bloom of toxic diatoms. For the first time, domoic acid was produced from these algae, and then ingested by marine life. Consumption of contaminated mussels resulted in 107 instances of amnesic shellfish poisoning, from domoic acid, including three deaths and permanent, short-term memory loss in several victims.

Also in 1987, there were major losses of sea urchin and coral communities in the Caribbean, a massive sea grass die-off near the Florida Keys, and on the beaches of the North Atlantic coast, the death of numerous dolphins and other sea mammals. It has been proposed that the combination of algal toxins, chlorinated hydrocarbons like PCBs, and warming may have lowered the immunity of organisms and altered the food supply for various forms of sea life, allowing *morbilli* (measles-like) viruses to take hold.

The 1990s

For five years and eight months, from 1990 to 1995, the Pacific Ocean persisted in the warm El Nino phase, which was most unusual, for since 1877 none of these distinctive warmings had lasted more than three years. Both anomalous phases—with either warmer (El Nino) or colder (La Nina) surface waters— bring climate extremes to many regions across the globe. With the ensuing cold (La Nina) phase of 1995-1996, many regions of the world that had lived with drought during the El Nino years were now besieged with intense rains and flooding. Just as in Colombia, flooding in southern Africa was accompanied by an upsurge of vector-borne diseases, including malaria. Other areas experienced a climatic switch of the opposite kind, with drought and wildfires replacing floods. During 1996 world grain stores fell to their lowest level since the 1930s. Weather always varies; but increased variability and rapid temperature fluctuations may be a chief characteristic of our changing climate system. And increased variability and weather volatility can have significant consequences for health and for society.

Decadal Variability

The cumulative meteorological and ecological impacts of the prolonged El Nino of the early 1990s have yet to be fully evaluated, and another is now upon us. In 1995, warming in the Caribbean produced coral bleaching for the first time in Belize, as sea surface temperatures surpassed the 29°C (84°F) threshold that may damage the animal and plant tissues that make up a coral reef. In 1997, Caribbean sea surface temperatures reached 34°C (93°F) off southern Belize, and coral bleaching was accompanied by large mortalities in starfish and other sea life. Coral diseases are now sweeping through the Caribbean, and diseases that perturb marine habitat, such as coral or sea grasses, can also affect the fish stocks for which these areas serve as nurseries. A pattern of greater weather variability has begun and is expected to persist with the El Nino of 1997 and 1998. Since 1976, such anomalies in Pacific Ocean temperatures and in weather extreme events have become more frequent, more intense, and longer lasting than in the preceding 100 years, as indicated in records kept since 1877.

DISCONTINUITIES AND INSTABILIBY

The common perception that the natural world changes only gradually can be misleading, for discontinuities abound. Animals switch abruptly between two states—awake and asleep—that are sharply divided and marked by qualitative differences in levels of activity in the central nervous system. Water can rapidly change from vapour to liquid to solid. Ecosystems have equilibrium states that are also at times abruptly interrupted. An extensive fire in an old growth forest, for example, can radically change the types of plants and animals within it.

Climate regimes can also change surprisingly fast. Recent analyses of Greenland ice cores indicate that significant shifts, called rapid climate change events (RCCEs), have taken place in the past in the span of but several years—not centuries, as was previously believed. While the oceans may serve as a buffer against sudden climate change, this mechanism may be limited, for some of the RCCEs seem to be associated with abrupt changes in ocean circulation.

The climate system exhibits equilibrium states as well, of which three may have been most common: when the poles of the Earth were covered with small, medium, or large ice caps. The present, Holocene period of the last 10,000 years—with medium-size caps and an average global temperature of

15°C (about 60°F)— has been associated with the development of modern agriculture and advancing civilization.

But our present climate regime may be becoming less stable. Increased variance—that is, more extreme swings—in natural systems is inversely related to how stable and balanced the systems are, and how sensitive they are to perturbations. Wider and wider variations can occur as a system moves away from its equilibrium state.

Trends in the 20th Century

The gradual warming that characterized the climate during the first four decades of the present century, for example, was accompanied by substantial temperature variability, as borne out in the record of degree-heating-days in the U.S. grain belt. The ensuing cooling trend from 1940 to the mid 1970s showed less variability. From 1976 to the present day, the variability—apparent in hot and cold spells, drought, and floods—has again increased. Greenland ice core records suggest that the last time the Earth warmed abruptly, ending the last Ice Age, there was also a pattern of increased variability.

The connection between human health and environmental stability increases our need for a better understanding of the present state of the global climate system. There are several unanswered questions regarding the system's stability. Was the drift toward earlier springtimes that began in this country in the 1940s indicative of the first minor readjustment in the climate regime? Are the more frequent and intense El Nino events since the mid 1970s another such indicator? Has the baseline of ocean temperatures shifted? Does the present climatic volatility—evident in altered weather and precipitation patterns—increase the potential for an abrupt "jump" in the climate system? And might further stresses lead to abrupt discontinuities of the type found in the Greenland ice cores, when the last Ice Age rapidly came to an end?

PROBLEMS ENCOUNTERED IN PUBLIC PARTICIPATION PRACTICE

The main areas of difficulty affecting public participation are attitudinal, lack of capacity to deliver programmes, lack of clarity about what outcomes are possible and the lack of a legislative framework. Some aspects of these issues are addressed below. The difference between success and failure is not always clear. If the public participation process delivers a clear decision, which is then overruled by the legislature, is this a success? A good example here is the Bantry Bay Integrated Coastal Zone Management (ICZM) Project, funded for three years under the EU Life Programme. After three years of intense community activity and the drawing up and enactment of a text-book public participation plan, the Bantry Bay Coastal Zone Charter was created.

At this point, no further funding was available, and the Project was closed down. Clearly the process was a success, but the result not. There is a general lack of experience of participatory processes in Ireland, and very few stakeholders have experienced planned participation processes. There may be suspicion, cynicism, or enthusiasm, but there is unlikely to be previous training. Suspicion about participation could be from politicians, who may feel their power is being diluted, or from ENGOs that have very specific and fixed viewpoints. Cynicism could be from previous public participation processes which were poorly managed, leaving participants reluctant to try again. Citizens are frustrated when they are treated as adversaries, rather than welcome participants in the EDM process........ they feel they have been invited too late in the process. Public 'hearings' often do not include 'listening'. Citizens efforts and ideas are not included in proposals and no reason given.

The resources available to proponents and government are overwhelming. Litigation and direct action at least get a reaction". Similarly, public hearings are criticised for: being held at times inconvenient for the public, establishing an atmosphere that inhibits dialogue, and conducting proceedings that intimidate the public. The lack of a clearly defined purpose can also means that the role of the public is sometimes unclear not only to the public but to those delivering the process, with the consequence that the participation process chosen is often unsuitable for the particular EDM process.

The making available of the necessary resources for a worthwhile public participation process, whether by the proponent or the regulator, by the commitment of personnel and finances, is essential. The implications for project proponents, of not having a properly planned and funded participatory EDM process, may prove much more costly, than dealing with the backlash. In many circumstances, the choice increasingly is not whether to involve the public, but how to get the best value from the chosen process. To avoid hearing from only the activist or the powerful elite, and in order to get the widest sweep of opinion and information, authorities must reach out into the community. In this regard some activities are best avoided, for example public meetings.

In the absence of trained facilitators these can often come down to 'he who shouts loudest wins!' Lack of technical support for the public, and difficulties in getting access to information can diminish the ability of the public to play a meaningful part in EDM processes. Public access to independent expertise in a particular technical/scientific field can prove difficult, particularly where the proponent of a project is a monopoly employer in the industry in the state. Faced with five weeks to object to a planning proposal, for example, this is a major hurdle. Added to this, it is the experience of the author that stakeholders with no technical/scientific background find it problematic to access, comprehend or evaluate data and information, especially under these short time constraints.

The feelings of powerlessness engendered by these hurdles, add to public perceptions of a lack of influence at both lower and higher levels of national and European government. "Consultations of interested parties.... can only ever supplement, and never replace, procedures and decisions of legislative bodies which possess democratic legitimacy". This sense of powerlessness may be part of the reason why, even in cases where considerable energy and resources are expended, to identify individuals and groups, only a small proportion of the public ever attend participation programmes. Public involvement may actually result in an increased level of conflict (134.p18). Becoming involved in a consultative process where the decision has already been made, or where the possible outcomes are not made clear at the outset, can lead to a great deal of frustration.

Anger may follow when the input of the public is ignored, especially following prolonged constructive engagement with an EDM process. Conflict can also occur between professionals. One source of such disagreements could evolve from the different emphases of 'scientific peer review' and 'social peer review'. The scientific peer review process is well established, and is essential to assess the technical information provided to policy makers. The parallel social peer review, designed to obtain societal acceptance and legitimacy for the decisions rendered, by contrast, does not have a set of recognised professional standards.. It should also be remembered that whilst science may strive for the 'truth', the opinions of scientists are coloured by their values and beliefs. Scientists, engineers and other 'experts' should show respect in the face of 'emotional', cultural and traditional forms of wisdom. The perceptions created by a technocratic philosophy encourage NIMBY activism and further polarise communities.

ENVIRONMENTAL IMPACT ASSESSMENT OF FORESTRY PROJECTS

The current Regulations are The Environmental Impact Assessment (Forestry) (England and Wales) Regulations 1999 [SI 1999/2228] and the Environmental Impact Assessment (Forestry) (Scotland) Regulations 1999 [SI 1999/43]. These Regulations have been further amended by the Environmental Impact Assessment (Forestry) (England and Wales) (Amendment) Regulations 2006 and The Environmental Impact Assessment (Scotland) Amendment Regulations 2006. These Regulations came into force on 6th September 1999 and require anyone who wishes to carry out a relevant project (i.e. deforestation, afforestation, forestry roads or quarries that might have a significant effect on the environment) to obtain consent for the work from the Forestry Commission. The applicant or proposer must submit an Environmental Statement in support of the proposals to apply for consent.

You will find below some of the terms used in the EIA Regulations:

a) Appropriate Authority-The Secretary of State for Environment, Food and Rural Affairs in England, the National Assembly for Wales, and the Scottish Executive in Scotland.

b) Countryside bodies
England-Natural England, Environment Agency. Wales-Countryside Council for Wales, Environment Agency. Scotland-Scottish Natural Heritage, Scottish Environmental Protection Agency. And any other body designated by statutory provision as having specific environmental responsibilities.

c) Determination-Taken from Regulation 15 "Determi-nation of applications" and is the process by which we make our decision about the application for consent.

d) Forestry projects-the project work that the Forestry Commission must assess under these Regulations are deforestation (conversion to another land use), afforestation, forestry roads and forestry quarries.

e) Opinion-Our consideration of the proposals from which we will decide whether or not the project is a relevant one. If it is, the applicant must apply for consent and provide an ES.

f) Relevant project-A forestry project (afforestation, deforestation, forest roads works and forest quarry works) that is likely, by virtue of factors such as its nature, size and location, to have a significant effect on the environment and as such requires the FC's consent.

g) Scoping-A gathering of all consultees and other interested parties to discuss and agree the significant issues of concern that require to be addressed by an applicant when preparing an Environmental Statement.

h) Screening-the process by which the Forestry Commission decides whether a project "is likely to have significant effects on the environment by virtue, inter alia, of its nature, size or location". This process is a distinct one from scoping.

i) Publicity-The Forestry Commission maintains a web-based EIA Register that gives details of all the decisions we make under these Regulations. Applicants are also required to advertise details of any application for consent.

j) Thresholds-Area limits set by the Regulations below which it is not expected that the project will have a significant effect on the environment.

The Forestry Commission will assess whether:

- o the proposed project is one of the following categories-afforestation, deforestation, forest roads or quarries;
- o the area is above the relevant threshold (includes extensions to similar areas of work);
- o the project is likely to have a significant effect on the environment.

Our consent will be required to proceed with the work if your proposals meet each of these requirements. You will be asked to provide an Environmental Statement to allow us to decide whether to give consent to the project. Under the 1999 Regulations, proposals are considered to be relevant projects (i.e. to require an EIA) if they fall within the categories listed below and the work proposed is likely to have a significant effect on the environment.

a) Initial afforestation: creating new woods and forests by planting trees (on an area that has not had trees for many years). This category includes using direct seeding or natural regeneration, planting Christmas trees and short rotation coppice;

b) Deforestation: conversion of woodland to another type of land use (e.g.heathland);

c) Forest roads: constructing forestry roads, including those within a forest and those leading to a forest;

d) Forest quarries: quarrying to obtain material (rock, sand and gravel) for the formation, alteration or maintenance of forest roads.

We may serve you with an Enforcement Notice if you carry out work on a project that would have required our consent. This notice will require you to comply with the Regulations. Situations may arise where one of the above forestry projects forms part of a wider development that requires Planning Permission.

In these circumstances, any necessary EIA will usually not be dealt with under the Forestry EIA Regulations but under the parallel Town and Country Planning EIA Regulations.

Where a project has been completed since 6 September 1999, work proposals of the same type (including that on land in different ownerships) that extend it beyond the thresholds may need our consent. This also applies where the accumulated area of all adjacent existing projects plus the new proposals, exceeds the thresholds.

PUBLIC PARTICIPATION IN ENVIRONMENTAL DECISION MAKING

The title of this study, poses several questions: What is the 'public'? What is 'participation'? What are the 'decisions' to which it refers? Starting with the last and working backwards, EDM refers to any process of decision-making where consequent significant environmental impacts are a possibility. This includes law making, planning, strategic planning, resource management planning, licensing of industry e.g. IPPC, environmental assessment (EIA), spatial planning etc. EDM can be even more complex than decision-making on other public issues.

First, environmental impacts do not respect property, jurisdiction or boundaries.

Second, EDM can involve government agencies as both manager and regulator.

Thirdly, environmental issues can provide especially heated value conflicts that require value trade offs.

The level at which the public is involved varies with the relevant legislation, and the attitude of the other stakeholders. Often it just means informing the public of a previously made decision and asking for comments, which may or may not be heeded. Sometimes it means informed consultation. Here there is exchange of information prior to the relevant authority's reasoned decision making, and all inputs are included and seen to be included. An example of this being the River Basin Management process under Article 14 of the Water Framework Directive as described by Judith Cuff (Cuff, J. 2001). Another example would be the Bantry Bay Coastal Zone Charter. Public participation, at its apex, could also mean that the public itself, in consultation with the relevant bodies, makes the final decision itself. Examples of this can be found under the auspices of Local Agenda 21. For public participation to be effective at any level, it requires the public to be well informed and kept aware of the possibility of participation. This requires a pro-active approach from industry and the relevant public bodies.

What then is 'the public'? The public is often treated as a unitary body, whereas in reality it is a collection of numerous continually shifting interests and alliances (Ortolano), which may be in conflict with each other. The term is used as a "catch-all to describe those with an interest in a decision, other than a proponent, operator, or responsible authority". (Petts and Leach). The individuals making up a public may be involved as individuals or as members of organisations. They may become involved due to their proximity, economics, social or environmental issues, values, etc. By contrast, stakeholders, of which the public is one, are literally those with a stake in an issue and may include non-governmental organizations (NGO's), government or its agents, industry, individuals, communities etc.

Stakeholders do not always want to be involved in an EDM process, but they have the right to know if their interests are affected. They may want to become involved at a different stake of the EDM process.

STATUS AND TRENDS OF DROUGHT AND DESERTIFICATION

Two thirds of Africa is classified as deserts or drylands. These are concentrated in the Sahelian region, the Horn of Africa and the Kalahari in the south. Africa is especially susceptible to land degradation and bears the greatest impact of drought and desertification. It is estimated that two-thirds of African land is already degraded to some degree and land degradation affects at least 485 million people or sixty-five percent of the entire African population. Desertification especially around the Sahara has been pointed out as one the potent symbols in Africa of the global environment crisis. Climate change is set to increase the area susceptible to drought, land degradation and desertification in the region.

Under a range of climate scenarios, it is projected that there will be an increase of 5-8 per cent of arid and semi Arid lands in Africa. Estimates from individual countries report increasing areas affected by or prone to desertification. It is estimated that 35 percent of the land area (about 83,489 km2 or 49 out of the 138 districts) of Ghana is prone to desertification, with the Upper East Region and the eastern part of the Northern Region facing the greatest hazards. Indeed a recent assessment indicates that the land area prone to desertification in the country has almost doubled during recent times. Desertification is said to be creeping at an estimated 20,000 hectares per year, with the attendant destruction of farmlands and livelihoods in the country.

Seventy percent of Ethiopia is reported to be prone to desertification, 26 while in Kenya, around 80 percent of the land surface is threatened by desertification. Estimates of the extent of land degradation within Swaziland suggest that between 49 and 78 per cent of the land is at risk, depending on the assessment methodology used (Government of Swaziland, 2000). Nigeria is reported to be losing 1,355 square miles (1mile =1.6km) of rangeland and cropland to desertification each year. This affects each of the 10 northern states of Nigeria. It is estimated that more than 30 per cent of the land area of Burundi, Rwanda, Burkina Faso, Lesotho and South Africa is severely or very severely degraded. These rates and extent of land degradation/desertification undermine and pose serious threats to livelihoods of millions of people struggling to edge out of poverty. They also cripple provision of land resources-based ecosystem services that are vital for a number of development sectors.

Drought is one of the most important climate-related disasters in Africa. Climate change is set to exacerbate occurrence of climate related disasters including drought. A study from Bristol University projects that areas of western Africa were at most risk from dwindling freshwater supplies and droughts as a result of rising temperatures.30 Current climate scenarios predict that the driest regions of the world will become even drier, 31 signalling a risk of persistence of drought in many parts of Africa (arid, semi-arid and dry sub humid areas) which will therefore bear greater and sustained negative impacts.

Impact of Drought and Desertification

It is common knowledge that land degradation and desertification constitutes major causes of 25. forced human migration and environmental refugees, deadly conflicts over the use of dwindling natural resources, food insecurity and starvation, destruction of critical habitats and loss of biological diversity, socio-economic instability and poverty and climatic variability through reduced carbon sequestration potential. The impacts of drought and desertification are among the most costly events and processes in Africa. The widespread poverty, the fact that a large share of Africa's economies depend on climate-sensitive sectors mainly rain fed agriculture, poor infrastructure, heavy disease burdens, high dependence on and unsustainable exploitation of natural resources, and conflicts render the continent especially vulnerable to impacts of drought and desertification.

The consequences are mostly borne by the poorest people and the Small Island Developing States (SIDS). In the region, women and children in particular, bear the greatest burden when land resources are degraded and when drought sets in. As result of the frequent droughts and desertification, Africa has continued to witness food insecurity including devastating famines, water scarcity, poor health, economic hardship and social and political unrest.32 The gravity of drought and desertification impacts in the region is demonstrated by the following examples.

CAUSES OF DROUGHT AND DESERTIFICATION

The underlying cause of most droughts can be related to changing weather patterns manifested through the excessive build up of heat on the earth's surface, meteorological changes which result in a reduction of rainfall, and reduced cloud cover, all of which results in greater evaporation rates. The resultant effects of drought are exacerbated by human activities such as deforestation, overgrasing and poor cropping methods, which reduce water retention of the soil, and improper soil conservation techniques, which lead to soil degradation. Desertification is caused by multiple direct and indirect factors. It occurs because drylands ecosystems are extremely vulnerable to over-exploitation and inappropriate land use that result in underdevelopment of economies and in entranced poverty among the affected population.

Whereas over cultivation, inappropriate agricultural practices, overgrasing and deforestation have been previously identified as the major causes of land degradation and desertification, it is in fact a result of much deeper underlying forces of socio-economic nature, such as poverty and total dependency on natural resources for survival by the poor. It is also true to reiterate that desertification problems are best understood within the dictates of disparities of income and access to or ownership of resources. Consequently, the causes of desertification are more complex to unravel. Desertification is driven by a group of core variables, most prominently climatic factors (Yang and Prince 2000; Hulme and Kelly 1993) that lead to reduced rainfall (Rowell et al. 1992) and human activities involving technological factors, institutional and policy factors, and economic factors (UNCCD 2004) in addition to population pressures, and land use patterns and practices. The technological factors include innovations such as the adoption of water pumps, boreholes, and dams.

The institutional and policy factors include agricultural growth policies such as land distribution and redistribution (AIBS 2004). These variables drive proximate causes of desertification such as the expansion of cropland and overgrasing, the extension of infrastructure, increased aridity, and wood extraction.Since most economies of African countries are mostly agro-based, a greater proportion of the desertification problems in rural areas are a result of poverty related agricultural practices and other land use systems. Inappropriate farming systems such as continuous cultivation without adding any supplements, overgrasing, poor land management practices, lack of soil and water conservation structures, and high incidence of indiscriminate bushfires lead to land degradation and aggravate the process of desertification.

These factors prevail in many parts of the region. In Uganda, as a result of overgrasing in its drylands known as the "cattle corridor," soil compaction, erosion and the emergence of low-value grass species and vegetation have subdued the land's productive capacity, leading to desertification. In the Gambia, it is reported that fallow periods have been reduced to zero on most arable lands. Between 1950 and 2006, the Nigerian livestock population grew from 6 million to 66 million, a 11-fold increase. The forage needs of livestock exceed the carrying capacity of its grasslands. It is reported that overgrasing and over-cultivating are converting 351,000 hectares of land into desert each year. The rates of land degradation are particularly acute when such farming practices are extended into agriculture on marginal lands such as arid and semi rid lands, hilly and mountainous areas and wetlands. Deforestation, especially to meet energy needs and expand agricultural land is another serious direct cause of desertification in the region.

Globally, there is evidence demonstrating a heavy negative impact of the energy sector on forest and other vegetation cover and land productivity. More than 15 million hectares of tropical forests are depleted or burnt every year in order to provide for small-scale agriculture or cattle ranching, or for use as fuel wood for heating and cooking. Biomass constitutes 30 percent of the energy used in Africa and over 80 percent used in many sub-Saharan countries such as Burundi (91 percent), Rwanda and Central Africa Republic (90 percent), Mozambique (89 percent), Burkina Faso (87 percent), Benin (86 percent), Madagascar and Niger (85 percent).8 Production and consumption of fuel wood is said to have doubled in the last 30 years of the 20th century and is rising by 0.5 percent every year. This high dependence on biomass fuel has resulted into an alarming rate of tree felling and deforestation, which is exposing large tracts of land to desertification. In Ghana, where the population density has reached 77 persons per km^2, 70 percent of the firewood and charcoal needed for domestic purposes comes from the savannah zones, as a result destroying 20,000 ha of woodland per annum. In Uganda where 90 percent of the population lives in rural areas and directly depends on land for cultivation and grasing, forestland shrank from 45 percent of the country's surface area to 21 percent between 1890 and 200011. In Nigeria where more than 70 per cent of the nation's population depends on fuel wood, it is feared that the country might be left with no forest by 2010 owing to the present level of deforestation activities. Already it is estimated that more than 13 million tonnes of soil are washed away into the sea annually. It is also feared that if the current rate of tropical forests deforestation is maintained, the tropical forests could be almost entirely harvested by the year 2050, thus devastatingly contributing to climate change, loss of biodiversity, land degradation and desertification. The above direct causes of desertification are driven by a complex set of underlying factors including the high levels of poverty in the region, high population growth rates, poor natural resources tenure and access regimes, conflicts, and climate change. Without alternatives poor people are forced to exploit land resources including fragile lands, for survival (food production, medicine, fuel, fodder, building materials and household items).

Given that most drylands in Africa are poverty hotspots as well, the risk of desertification is high in many of these areas, as the poor inevitably become both the victims and willing agents of environmental damage and desertification. In Sub-Saharan Africa alone 270 million people live in absolute poverty. In Uganda, over 40 percent of the pastoralists who constitute the majority in the country's drylands, live below the poverty line. High population growth increases pressure on limited and fragile land resources.

The rural population living in drylands in Africa is estimated to be 325 million. This breeds favourable conditions for deforestation and overexploitation of land that lead to land degradation as a large and growing rural population, struggling to survive in a limited natural resource base result in the over-utilisation of the available natural resources. For instance the Nigeria's human population which grew from 33 million in 1950 to 134 million in 2006, a fourfold expansion has forced farmers to plough marginal land under the pressure to meet food needs. As a result of this, the country is slowly turning into a desert. According to the New York Times, Niger's population has doubled in the last 20 years. Each woman bears about seven children, giving the country one of the highest growth rates in the world. Given that 90 percent of Niger's people live off agriculture, this population is exerting great pressure on the less than 12 percent of its land that can be cultivated. Insecure and unclear land and other natural resources tenure and access rights are some of the main reasons the natural resources end-users are unwilling to invest in long-term sustainable land management (SLM). For instance it is reported that in Uganda, insecurity of land tenure in parts of the cattle corridor under mailo and communal land ownership systems does not encourage farmers to invest in sustainable land management practices.

Bibliography

Alka Pareek: *Environment and Nutritional Disorders*, Aavishkar Pub, Delhi, 2003.

Allen, P.: *Food For the Future: Conditions and Contradictions of Sustainability*, New York: John Wiley and Sons, 1993.

Bahati, J.: *Impact of Arboricidal Treatments on the Natural Regeneration of Species Composition in Budongo Forest*, Uganda. M. Sc., Makerere University, 1997.

Banwari Lal: *Environmental Microbiology*, Cyber Tech Pub, Delhi, 2009.

Barlow K: *Ecology of Food and Nutrition*, Oceania, New Guinea, 1984.

Bethell, E.: *Vigilance in Foraging Chimpanzees*. University College, London University, 1998.

Bhatia, S C : *Handbook of Environmental Microbiology*, Atlantic, Delhi, 2008.

Biswas, A., and Cline, S.: *Global Change: Impacts on Water and Food Security*, Springer, Heidelberg, 2010.

Carley, M., P. Jenkins : *Urban Development and Civil Society, the Role of Communities in Sustainable Cities*, Sterling, VA, Earthscan Publications, 2001.

Casado, Matt A: *Food and Beverage Service Manual*. New York: Wiley, 1994.

Chanderlekha Goswami: *Biochemistry and Instrumentation*, Manglam Pub, Delhi, 2011.

Chouhan, T.S.: *Desertification in the World and Its Control*, Scientific Publishers, Delhi, 1992.

Florkin, M.: *Comprehensive Biochemistry: Amsterdam*, Elsevier, 1975.

Fumento, Michael: *Bioevolution: How Biotechnology is Changing Our World*, San Francisco, Encounter Books, 2003.

Garg, G.P. : *Biochemistry of Nucleic Acids*, Discovery Pub, Delhi, 2010.

Gay, Kathlyn: *Saving the Environment: Debating the Costs*, New York, Franklin Watts, 1996.

Geist, Helmut: *The Causes and Progression of Desertification*, Ashgate Publishing, Delhi, 2005.

Goel, A K : *Basic Concept of Animal Chemistry*, Pearl Books, Delhi, 2008.

Goodman, D.C.: *From Farming to Biotechnology: A Theory of Agro-industrial Development*, Oxford, Blackwell, 1987.

Govind Prasad: *Environmental Geomorphology*, Discovery Publishing House, Delhi, 2008.

Govindan, R. : *Environmental Microbiology and Instrumentation*, Manglam, Delhi, 2009.

Gunjan Goel : *Applied Dairy and Food Microbiology*, Agrotech, 2005.

Harsh Bhaskar: *Problems in Biochemistry*, Campus Books, Delhi, 2010.

Jack Simons: *An Introduction to Theoretical Chemistry*, Cambridge Univ Press, Delhi, 1998.

Jai Shankar Ojha: *Aquaculture and Nutrition and Biochemistry*, Agrotech, Delhi, 2006.

Jasra, O.P. : *Environmental Biochemistry*, Sarup & Sons, Delhi, 2002.

Katz, P., V. J. Scully : *The New Urbanism, Toward an Architecture of Community*, New York, McGraw-Hill, 1994.

Mahboob, Syed : *Handbook of Fruit and Vegetable Products*, Agrobios, Delhi, 2008.

Meyer, Art: *Earth Keepers: Environmental Perspectives on Hunger, Poverty, and Injustice*, Scottsdale, Herald Press, 1991.

Nelson DL, Cox MM: *Lehninger's Principles of Biochemistry*, New York, New York: W. H. Freeman and Company, 2005.

Oliver, John E.: *Encyclopedia of World Climatology*, Springer, Delhi, 2005.

Perrings, C. : *Sustainable Development and Poverty Alleviation in Sub-Saharan Africa*, Botswana, New York, St. Martin's Press, 1996.

Pollock NJ: *The Concept of Food in Pacific Society: A Fijian Example*, Oceania, Fiji, 1985.

Prasad, S.K. : *Biochemistry of Proteins*, Discovery, Delhi, 2010.

Ram Naresh Mahaling: *Basics of Biochemistry*, Anmol, Delhi, 2008.

Sarath Chandra Bose: *Biochemistry : A Practical Manual*, Pharma Med Press, Delhi, 2010.

Shiel, D.: *The Ecology of Long-term Change in a Ugandan Rainforest*. D. Phil., Oxford University, 1997.

Shubhrata R. Mishra: *Plant Biochemistry*, Discovery, Delhi, 2010.

Simonds, J. O. : *Garden Cities 21, Creating a Livable Urban Environment*, New York, McGraw-Hill, 1994.

Singh, Mahinder: *A Textbook of Biochemistry*, Dominant Pub, Delhi, 2011.

Sobsey M (2002) *Managing food in the home: Accelerated health gains from improved water supply.* Geneva, World Health Organization.

Suri, Nitin : *Molecular Biology and Biochemistry*, Oxford Book Company, Delhi, 2010.

Tucker, Mary Evelyn, and John Grim: *Worldviews and Ecology: Religion, Philosophy, and the Environment.* Orbis Books, Maryknoll, N.Y. 1994.

Uhlir, P. F. : *Scientific Data for Decision Making Toward Sustainable Development*, Washington, D.C., National Academies Press, 2003.

Ullrich M *Bacterial Polysaccharides: Current Innovations and Future Trends* Caister Academic Press, 2009.

Wade, J. L.: *Society and Environment: The Coming Collision*, Boston, Allyn and Bacon, 1972.

Walter, B., L. Arkin : *Sustainable Cities, Concepts and Strategies for Eco-city Development*, Los Angeles, CA, EHM Eco-Home Media, 1992.

Webb, R. : *Floods, Droughts and Climate Change*, University of Arizona Press, NY, 2001.

Zupan, J.M.: *The Distribution of Air Quality in the New York Region*, Baltimore, Johns Hopkins University Press, 1973.

Index

❑❑❑